A Naturalist's Guide to the

TREES & SHRUBS OF INDIA

Pakistan [illegible] Bangladesh [illegible] Lanka

Pradeep Sachdeva and Vidya Tongbram

JOHN BEAUFOY PUBLISHING

First published in the United Kingdom in 2017 by John Beaufoy Publishing Ltd
11 Blenheim Court, 316 Woodstock Road, Oxford OX2 7NS, England
www.johnbeaufoy.com

10 9 8 7 6 5 4 3 2 1

Photo Credits
Front Cover: Clockwise from left Mango, Laburnum, Flame of the forest, White Gulmohur, Barna (Pradeep Sachdeva and Vidya Tongbram)
Back Cover: Canonball Tree (Pradeep Sachdeva). **Title page:** Queens Crape Myrtle (Pradeep Sachdeva)
Contents Page: White Bauhinia (Pradeep Sachdeva)

All photographs by Pradeep Sachdeva and Vidya Tongbram except **Arthur Duff** 5, **Mimi Roberts** 31b, **Alpana Khare** 35b, **Pradip Krishen** 52, 125b, **Nicholas Vreeland** 53t, **Gowri Mohanakrishnan** 75b, and **CSE-Down to Earth** 83.

ISBN 978-1-909612-82-2

Edited by Krystyna Mayer
Designed by Gulmohur India

Printed and bound in Malaysia by Times Offset (M) Sdn. Bhd.

Contents

INTRODUCTION

Trees and shrubs play a vital role in maintaining an ecology that nurtures life and makes it habitable. Their role in our environment is essential. In their most basic process of photosynthesis, plants absorb carbon dioxide and produce oxygen that is crucial for the survival of all life forms. Humans with their innovative nature have furthered this dependence on plants by creating ways of using timber to build shelters, formulating brews that are used medicinally, weaving fabrics out of their fibre for clothing and producing high-yielding crops for food and so on. The ways in which people depend on plants are numerous across the world.

In India, this dependence is recognized and celebrated by attaching special cultural and religious significance to plants. For example, the Peepal, referred to as the Bodhi Tree, is the tree under which Buddha achieved enlightenment. We can see this in the friezes and sculptures of the Buddhist Stupas that record the worship of the Bodhi Tree as a symbol of enlightenment and as an iconic representation of Buddha himself.

Such portrayals of the significance of trees are consistent historically, across time and faiths, as is evident in various Mughal tombs and mosques where the botanical patterns appear repeatedly.

On a domestic, day-to-day front, Tulsi or Sacred Basil is venerated and planted in the forecourt or central courtyard of Hindu households. It is regarded as a plant of high

Frieze showing worship of the Bodhi tree along with the empty throne at Bharhut Stupa, Madhya Pradesh and Amaravati.

Tree and palm motif of stone lattice at Siddi Sayyed Mosque, Gujarat.

medicinal value and also used in religious ceremonies. One of the age-old practices of using various plants has been in traditional medicine. The Mango is another celebrated plant, with mango festivals being organized in summer showcasing the varieties of species available. The leaves of Mango are also associated with auspicious symbolism and used as decorative motifs in ceremonies

Most Hindu and Sikh temples, and Muslim dargahs have been associated with sacred trees. The Bishnoi community of the Rajasthan desert regards the Khejri (*Prosopis cineraria*) as its sacred tree. An entire movement to protect trees from illegal felling, Chipko (hug a tree) was started in the North Indian Himalaya. Sanctification of trees has been mentioned in various ancient texts such as the *Rig Veda*, *Atharva Veda*, *Mahabharata* and *Ramayana*, which is perhaps a way of communicating to people about protecting trees. These ancient texts have also helped scholars to locate the origins and importance of particular trees. For instance, in *Ramayana*, Valmiki writes about edible and non-edible vegetation, sacred and medicinal plants, and forest ecosystems.

Increasingly plants, especially those grown in forests, are being valued as a resource for mitigating global warming because of their capacity to absorb carbon dioxide and generate oxygen. This is most relevant in today's context of growing urban sprawls, due to which forests cover and urban parks are shrinking.

PLANT HABITATS OF INDIA

The regions of India are diverse and support a large range of plants. There are at least 18,000 different flowering plants in India and this book covers some of the most common trees and shrubs found in India, as well as Nepal, Bangladesh, Pakistan and Sri Lanka.

The plant habitat in India is very rich and diverse, embracing different ecosystems. It ranges from the hot, arid desert regions of Rajasthan to the fertile areas of the Indo-Gangetic plains. It also nurtures the conifers and broadleaved plants of the Eastern and Western Himalaya to the dry scrub of the Punjab. The tropical green rainforests extend from North-east India to Kerala in the south. Andaman and Nicobar Islands in the Bay of Bengal are a distinct landscape similar to South-east Asia.

Many plants are native to the region and species were introduced over millennia by traders, and colonisers such as the Portuguese, British and Mughals. Below are the various identified regions that spread across India, providing different habitats:

Western Himalaya Evergreen and deciduous broadleaved forests stretching from Kashmir to Kumaon to West Nepal.
Eastern Himalaya It is a tapestry of temperate subalpine coniferous forests, subtropical jungle, savannah and rich alpine meadows stretching through North Bengal, Sikkim to Arunachal Pradesh.
North-west Dry Region Thar desert, stretching from Punjab, Haryana, Rajasthan and Gujarat.
Gangetic Plains Stretching from plains of west Uttar Pradesh to Hooghly delta and forests of Sunderbans.
The Western Ghats Home to four tropical and subtropical moist broadleaved forest ecoregions – the North Western Ghats moist deciduous forests, North Western Ghats

Sal forest.

rainforests, South Western Ghats moist deciduous forests, and South Western Ghats rain forests. Stretch from South Gujarat to Kanyakumari.

Central Indian Region Satpura range, tropical thorn forests and mangroves on delta of Godavari.

Deccan and Carnatic Central Deccan Plateau dry deciduous forests. This forest ecoregion of southern India stretches from Peninsular India south of the Godavari, to Maharashtra, Andhra Pradesh, Karnataka and Tamil Nadu.

North-east India Evergreen rainforests, tropical semi-evergreen forests and subtropical hill forests.

Andaman and Nicobar Islands Mangroves and wetlands, evergreen forests, semi evergreen, deciduous forests.

Palm-lined backwaters of Kerala.

Cedar forest, Himachal Pradesh.

IDENTIFYING PLANTS

This guide is aimed at giving the reader a true and correct description of plants in order to identify them in the field. This is an effort that is sometimes precarious because plants, even of the same species, have varying growth habits according to the climate and soil conditions. Heights of trees can vary significantly from location to location.

Certain features such as the leaf, flower, and fruit, however, remain constant. Since no plant is permanently in flower or fruit, it is the leaf that often seals the process of identification.

At a distance the first sign of recognition is the form, and this is true especially of trees that have distinctly identifiable forms as compared to shrubs, which are often pruned by the gardener or foraged by cattle.

Horizontal spread of the branches of Indian Almond.

Dense, rounded crown of Maulsari.

Tapering form of Pine.

Open, spreading crown of Rain tree.

On approach, the shape of the trunk and texture of the bark help in giving a clear indication of the species in question, such as the pale, flaky bark of a eucalyptus tree that is distinctive in identifying the tree. The shape of trunks can vary from species to species. They can be straight boles like those of a poplar or crooked and gnarled like those of a Dhak.

The texture of the bark ranges from smooth to flaky or fissured, and is sometimes distinctly spiky like that of a young *Chorisia*, which when mature becomes less spiky.

Fissured bark of a mature caper.

Fissured, green-striped bark of young Chorisia.

Bark of Cupressus.

Flaky bark of eucalyptus.

On closer inspection, the factors that confirm the species of a plant are the leaf, flower and sometimes its distinctive fruit.

Leaves are classified as simple or compound.

Simple leaves are often identified by their shape and arrangement. For example, a Peepal leaf is heart shaped with an acutely pointed tip.

Simple leaf of Mango tree, shaped like a lance.

Simple leaf of the Peepal tree is heart shaped with an acutely pointed tip.

Simple leaf of the Mulberry tree, in which young leaves are lobed and have varying shapes that morph as they mature.

Simple leaf of Jackfruit tree, arranged alternately.

Simple leaf of Plumeria, *arranged in whorls at the tip of the branch.*

Compound leaves are easier to identify and are common to certain groups of trees. For example, the palmately compound leaf defines most trees of the mallow family. They are further sub-classified as palmate, pinnate or feathered, twice pinnate, or rarely thrice pinnate. There are also compound leaves made up of two leaflets like those of Anjan and *Bauhinia* species and sometimes three leaves as in Barna and Flame of the Forest.

Compound leaves of Bauhinia*, comprising two leaflets.*

Palmate leaf of Chorisia, 5–7 *leaflets joined like the fingers on the palm of a hand.*

Pinnate leaf of Neem and Tamarind trees with leaflets arranged in opposite pairs.

Thrice pinnate leaf of Drumstick tree.

Most plants can be clearly identified by their flowers. A most distinct example is the flower of the Bird of Paradise whose bird-like shape and dramatic colours leave no doubt to its species. However some flowers are inconspicuous in shape, such as that of Maulsari, but have a distinct smell that identifies the tree.

Edible fruits such as apples, pears, and jackfruits require little description and are easily recognized by most people. Some fruits, however, are not so common, but easily identify the plant, such as the rattling pods of Siris or the long, bean-like pods of Babool with a curvy edge as shown below.

Cone-like fruit of magnolias.

Flat, woody capsule of Harsingar.

Long pods of Siris, which remain on the tree and rattle with the wind.

Bean-like pods of Babool, resembling a string of beads.

SACRED GROVES OF INDIA

Many communities in India practise nature worship and certain groves of trees have special significance and are considered sacred. A manifestation of this worship is protection of patches of forests, dedicating them to ancient spirits and deities. These are known as sacred groves and can nurture single or multiple species of tree.

Their distribution is widespread in India with the highest concentration in the south-central part of the country.

Access to sacred groves is generally restricted and therefore human impact in those areas has been minimal. This has lead to them being important pools of biological diversity and allowed the complex ecological process to remain undisturbed. They have also become important resources for water due to the preservation of the springs, ponds and lakes within them.

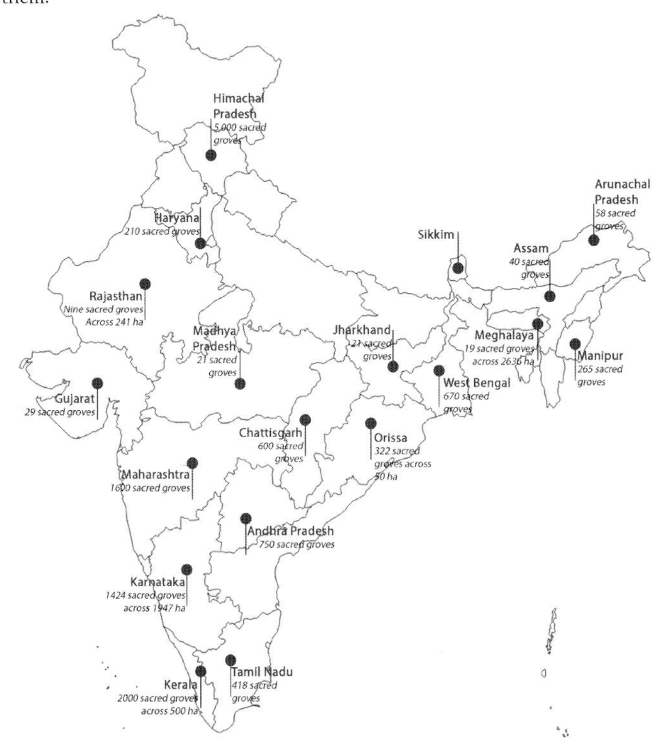

Map showing location and density of sacred groves across India. Data courtesy CSE - Down to Earth.

CLASSIFICATION AND FAMILY

For the purpose of easy reference, the plants have been grouped in alphabetical order by their family names. In order to simplify the identification and groups, the scientific name of each family has been further simplified to a common name. For example, plants of the Rosaceae family have been simply classified under Rose.

No.	Group	Family Name
1	Acanthus	Acanthaceae
2	Agave	Agavaceae
3	Araucaria	Araucariaceae
4	Asparagus	Asparagaceae
5	Bald Cypress	Taxodiaceae
6	Barberry	Berberidaceae
7	Beech	Fagaceae
8	Bixa	Bixaceae
9	Bignonia	Bignoniaceae
10	Borage	Boraginaceae
11	Brazil Nut	Lecythidaceae
12	Buckthorn	Rhamnaceae
13	Cannabis	Cannabaceae
14	Canna	Cannaceae
15	Caper	Capparaceae
16	Cashew	Anacardiaceae
17	Castor	Phyllanthaceae
18	Citrus	Rutaceae
19	Coffee	Rubiaceae
20	Combretum	Combretaceae
21	Costus	Costaceae
22	Crape Myrtle	Lythraceae
23	Custard Apple	Annonaceae
24	Dillenia	Dilleniaceae
25	Dipterocarp	Dipterocarpaceae
26	Dogbane	Apocynaceae
27	Euphorbias	Euphorbiaceae
28	Fig	Moraceae
29	Frankincense	Burseraceae
30	Garcinia	Guttiferae
31	Ginger	Zingiberaceae
32	Heath	Ericaceae
33	Horseradish tree	Moringaceae
34	Hydrangea	Hydrangeaceae
35	Laurel	Lauraceae
36	Magnolia	Magnoliaceae
37	Mahogany	Meliaceae
38	Mallow	Malvaceae
39	Mint	Lamiaceae
40	Myrtle	Myrtaceae
41	Nightshade	Solanaceae
42	Oleaster	Eleaegnaceae
43	Olive	Oleaceae
44	Oxalis	Oxalidaceae
45	Pandanus	Pandanaceae
46	Pea - Caesalpinea	Fabaceae/ Caesalpinioidae
47	Pea - Mimosa	Fabaceae/ Mimosoideae
48	Pea - Papilon	Fabaceae/ Papilionoideae
49	Pine	Pinaceae
50	Plane tree	Platanaceae
51	Plumbago	Plumbaginaceae
52	Poppy	Papaveraceae
53	Rose	Rosaceae
54	Tea	Theaceae
55	Tree of Heaven	Simaroubaceae
56	Rose	Rosaceae
57	Sapodilla	Sapotaceae
58	Sandal Wood	Santalaceae
59	Salvadora	Salvadoraceae
60	Sterlitzia	Strelitziaceae
61	Soap Berry	Sapindaceae
62	Silky Oak	Proteaceae
63	Verbena	Verbenaceae
64	Willow	Salicaceae

GLOSSARY

acute Sharply pointed.
alternate Not opposite.
angiosperms Flowering plants; plants with developing seeds enclosed in an ovary.
anther Part of the stamen containing pollen.
aril Membranous or fleshy appendage that partly or wholly covers a seed, for example, the fleshy outer layer of lychee fruit.
axil Point between leaf and stem from which buds emerge.
berry Fleshy, soft-coated fruit containing several seeds.
bract Modified leaf growing at the base of a flower or inflorescence.
capsule Dry fruit pod that splits and contains seeds.
compound Leaf that is divided into a number of leaflets.
corolla The petals of a flower, together.
corymb Flat-topped cluster of flowers.
cone Woody, seed-bearing structures found on conifers such as pines and firs.
cultivar Plant species created by cultivation.
deciduous Plants that shed leaves in certain seasons.
frond Feathery leaf of a fern or palm.
fruits Seeds of a plant.
hybrid Plants developed from cross-fertilization of two species.
inflorescence Flowering structure including bracts.
lanceolate Narrow and lance-shaped leaf.
linear Slender and parallel sided.
native Occurring naturally in a region.
mangrove Shrub or small tree growing in salt or brackish water, usually characterized by pneumatophores. Tropical coastal vegetation is characterized by such species.
midrib Central vein of a leaf.
ovate Egg-shaped leaf, bract or petal.
palmate Leaf shaped like an open hand.
petal In a flower, one of the segments or divisions of the inner whorl of non-fertile parts surrounding the fertile organs, usually soft and conspicuously coloured.
petiole Leaf stalk.
pinnate Compound leaf with leaflets arranged on each side of a common petiole or axis; also applied to how the lateral veins are arranged in relation to the main vein.
pod Elongated fruit, often cylindrical.
pollen Tiny grains that contain male sex cells produced by a flower's anthers.
raceme An inflorescence in which the youngest flowers are at the top.
sepal In a flower, one of the segments or divisions of the outer whorl of non-fertile parts surrounding the fertile organs, usually green.
stamen Male part of flower, comprising anther and filament.
stellate Star-like.
stigma Top of female part of flower to which pollen is deposited.
whorl Several leaves or branches arising from same point on a stem.

Vasaka *Adhatoda vasica*

Shrub. Height: 1–3m

DESCRIPTION Large, evergreen shrub with dark green leaves each shaped like a lance. White flowers are arranged on a spike. A variation with red flowers and darker leaves is also found. **HABITAT** Native to India; distributed widely in the plains and grows up to the foothills of the Sub-Himalaya. Often found on wasteland. **USES** Cultivated as a herbal plant – important medicinal plant in Ayurvedic, Siddha and Unani systems of medicine. Root, leaves and flowers are used to treat respiratory conditions, coughs and fever, and also to stop bleeding. The leaves are rich in vitamin C and yield an essential oil.

White Barleria

Barleria cristata

Shrub. Height: 0.5–1.5m

DESCRIPTION Evergreen shrub with simple leaves. They are lance shaped, covered in fine hair and arranged in opposite pairs. Trumpet-shaped, attractive white or blue flowers bloom in summer. Belongs to large genus of about 250 species. **HABITAT** Native to India and can be found across the country. **USES** Popular as a landscape plant. Can also be used as a bedding plant in partially shady spots, though it is happiest in full sun. Leaves and roots are used to treat coughs and inflammations.

Vajradanti

Barleria prionitis
Shrub. Height: 0.5–1.5m

DESCRIPTION Evergreen, hardy and spiny shrub with elliptical leaves. Yellow flowers are trumpet shaped and bloom in summer. **HABITAT** Native to India and can be found growing wild in dry, difficult conditions. Can grow in rocky areas and favours well-drained soil. Listed as an invasive weed, and known to take over native vegetation. **USES** A hardy plant that is suited to regenerated wastelands. Leaves can be chewed to relieve toothache. Paste of the root is applied on skin to reduce swelling.

Crossandra

Crossandra undulaefolia
Shrub. Height: up to 1m

DESCRIPTION Small, evergreen shrub with compact growth and attractive blossoms. Dark green leaves are simple, arranged in opposite pairs and have a wavy margin. Flowers appear in spikes nearly all year round and are profuse in warm months. They are colourful and showy, five lobed, in various colours ranging from yellow, orange and red, to pink. **HABITAT** Native to India and Africa, and grows across most parts of the Indian plains. Does well in semi-shade and in sandy, well-drained soil. **USES** Popular as a bedding plant in gardens for its colourful blossoms.

Blue Sage *Eranthemum nervosum*

Shrub. Height: 1m

DESCRIPTION Hardy, spreading shrub with dense foliage. Dark green leaves are simple, arranged in opposite pairs and have prominent veins. They are large, elliptical in shape and have pointed ends. Flowers are tubular, purple-blue, and borne on spikes in winter, in January–April. **HABITAT** Native to India and Sri Lanka, this is a plant suited to tropical and subtropical climates. Widely cultivated in Indian gardens as an ornamental shrub. Prefers full sun but also does well in partial shade. It is susceptible to frost. **USES** Suited for use as a bedding plant, especially to add colour to semi-shaded areas of the garden.

King's Mantle *Thunbergia erecta*

Shrub. Height: up to 2m

DESCRIPTION Sprawling shrub from a family that largely comprises climbers. The numerous branches shoot upright and are quick growing. Dark green leaves have acute, pointed tips and are shed in winter. Flowers are borne alone or in pairs, appearing several times a year in spring and during monsoons. They are trumpet shaped and deep purple in colour with yellow throats. A while flowering variety, var. *alba*, is also found. **HABITAT** Native to tropical Africa, and widely grown in India, it can tolerate varying soil conditions from sandy to clayey and acidic to alkaline. Thrives in full sun and also does well in partial shade. **USES** Widely cultivated in Indian gardens and suited to being pruned and maintained as a hedge.

Lipstick Tree ■ *Bixa orellana*

Shrub. Height: 4–5m

DESCRIPTION Dense, evergreen shrub with large green leaves. Heart-shaped leaves have acutely pointed tips. Flowers appear in clusters of pale pink blossoms, and are followed by fruits in a spiky pod that are red-brown in colour. They ripen to reveal seeds covered in a red aril, which is used to produce a dye. **HABITAT** Native to tropical America, and cultivated widely for its fruits and as a specimen plant. **USES** Dye used as a colouring in food and cosmetics, especially in lipsticks, from which it gets its name. Shrub also sometimes trained as a hedge plant. Most parts of the plant have medicinal properties, and it is used in local herbal remedies in its native region.

Agave ■ *Agave attenuata*

Shrub. Height: up to 2m

DESCRIPTION Evergreen shrub with succulent leaves that are acutely pointed. Colour of leaves varies from light grey to yellow-green. Inflorescence is a long raceme up to 2.5m high with greenish-yellow flowers is dense clusters. Due to the curved stem of the inflorescence, it is sometimes referred to as Swan's Neck or Lion's Tail. **HABITAT** Agave is from the Caribbean region of Mexico and the West Indies. This is not the most hardy of agave species and requires a moist environment. It thrives in partial shade.

New Caledonian Pine

■ *Araucaria columnaris*

Tree. Height: up to 30m

DESCRIPTION Tall, evergreen conifer which is considered an ancient tree species. It has a column-like form with a tapering crown and a formal structure. Its branches extend out horizontally. Dark green leaves have a stiff and leathery texture, and are arranged in spirals. Both male and female cones appear on the same tree. The female cones are larger than the male ones and take a couple of years to mature. **HABITAT** Tree of coastal areas endemic to islands of New Caledonia and Polynesia in the South Pacific area. Widely planted in India in inland areas as an ornamental tree. It prefers moist growing conditions in full sun. **USES** Tree is decorated at Christmas and is popularly referred to as the Christmas tree, although the tree originally used at Christmas is the Norfolk Pine. It is often grown as an accent tree in gardens or in rows along an avenue.

Dracaena ■ *Dracaena fragrans*

Shrub. Height: up to 5m

DESCRIPTION Evergreen shrub that can grow tall and narrow. Mature plants develop branches. Leaves are dark green, long and shaped like a lance. They are stalkless and grow spirally directly from the stem. Flowers are fragrant and small, clustered along a spike. Fruits are berry-like and yellow-orange. Variegated cultivars 'massangeana' and 'victoriae' with striped leaves are also found. **HABITAT** Introduced in India from tropical parts of Africa. It is widely grown in Indian home gardens and prefers a location with partial shade. **USES** A popular houseplant, it is commonly grown in containers as an indoor plant.

Song of India ■ *Dracaena reflexa 'variegata'*

Shrub. Height: up to 4m

DESCRIPTION Attractive foliage plant, mostly cultivated as a houseplant in containers. Can grow to a large shrub in the ground. Evergreen leaves are dark green in colour. It is long and narrow, shaped like a lance and grow out spirally from the stem. A popular variety in Indian gardens has variegated leaves with streaks of yellow, and is commonly known as the Song of India. **HABITAT** Native to Madagascar, Mauritius and parts of western India. Popular houseplant that does well in partial shade. It prefers moist but well-drained growing conditions. **USES** An ornamental shrub, often planted in pots indoors.

Furcraea ■ *Furcraea foetida*

Shrub. Height: 1–2m

DESCRIPTION Spiky plant whose thick, succulent leaves are sharp and grow upright in rosettes. It has no stem and resembles an agave plant. Flowers are creamy-white and small, and grow in clusters on tall spikes that reach up to several metres in height. Variegated cultivar with yellow streaks is also popular in cultivation. **HABITAT** Native to northern parts of South America, and widely grown in most parts of India. **USES** Hardy accent shrub grown in roadside landscapes and gardens. Natural fibre from the leaves is used for making ropes and coarse fabrics.

Yucca ■ *Yucca aloifolia*

Shrub. Height: 1–2m

DESCRIPTION Evergreen tropical plant with erect trunk covered with leaves. These are stiff, long and narrow, with sharp, pointed ends. The leaves resembles swords and the plant is also known as the Spanish Bayonet. With age, the leaves drop to reveal a long, bare stem. Flowers are white and showy, and grow in large clusters on a spike. **HABITAT** Introduced from parts of North America and Mexico. Hardy plant that can adapt to various soil conditions and does well in semi-arid regions. **USES** Planted as an ornamental shrub in gardens, where it is a popular accent plant.

Montezuma cypress

■ *Taxodium mucronatum*

Tree. Height: up to 40m

DESCRIPTION Large tree known for its longevity. It is deciduous in cold climates. Needle-like leaves are densely arranged on drooping branches. Bright green leaves turn reddish-brown in autumn before shedding. Male and female cones appear on same tree. Minute flowers appear in strings in spring. **HABITAT** Native to Mexico; grows along riversides and streams. Suited to well-drained soil and can withstand prolonged flooding. **USES** Wood is resistant to rot and decay, and is used in general construction. Bark, roots and leaves have medicinal value.

Heavenly Bamboo

■ *Nandina domestica*
Shrub. Height: up to 2m

DESCRIPTION This is not a bamboo, as suggested by its name, but a bushy shrub with erect stems. Leaves are evergreen and change colour to a purple tinge in autumn and winter. They are compound, being twice or three times pinnate. Flowers are small and white, and are produced in clusters followed by red berries. **HABITAT** Native to China and Japan. Favours semi-shaded, moist locations. **USES** Cultivated in Indian gardens as an ornamental shrub.

Banjh Oak

■ *Quercus leucotrichophora*
Tree. Height: up to 30m

DESCRIPTION Evergreen tree that grows tall and is known to live for a very long time. Lance-shaped leaves have serrated edges, are silver-grey underneath and are arranged alternately. Flowers are minute, appearing in spikes. Male catkins are borne on branch tips and female flowers grow on the bases of leaves. Fruits are oval-shaped nuts, called acorns. **HABITAT** Grows at high altitudes in the Himalayan range, though a few trees grow and thrive in a place as low as Dehradun. Prefers mild and moist climates. **USES** Wood has a tendency to warp and spilt, and is used in building construction and for making agricultural tools. Wood pulp is used to make hardboards. Acorns possess medicinal properties that are astringent and diuretic. Local communities use the timber as fuel wood, and the leaves as fodder for their cattle.

Kamandal Tree ■ *Crescentia cujete*

Tree. Height: 6–10m

DESCRIPTION Small tree with unruly branching pattern. The sparse foliage is semi-deciduous and sheds in winter. Simple leaves are oval with rounded ends, dark green and leathery. The bell-shaped flowers are greenish-yellow with purple veins, and are borne directly on trunks and old branches. Fruits are gourd-like with a hard woody cover, containing numerous seeds in a pulp. **HABITAT** A tree from Central America and northern parts of South America, it is not common in India and is planted as a specimen tree. **USES** The fruit is cut in half and the hard cover of the gourd is polished and used as a container. In India, *Sadhus* are known to carry begging bowls made of calabash fruits called *Kamandal*.

Neeli Gulmohur

■ *Jacaranda mimosifolia*
Tree. Height: 10–15m

DESCRIPTION Ornamental tree with a light, feathery crown. Leaves are compound and feather-like, and composed of small leaflets arranged in opposite pairs. Mauve-purple flowers appear in clusters, shaped like a trumpet, in March–May. Fruit is a rounded, flat, woody pod containing a winged seed. **HABITAT** Introduced to India from South America. In India cultivated widely as an ornamental. **USES** Jacaranda trees are planted in parks and gardens as ornamental trees.

Sausage Tree

■ *Kigelia pinnata*
Tree. Height: up to 15m

DESCRIPTION Quick-growing, evergreen tree of medium size that is deciduous in dry conditions. Compound leaves have oblong leaflets arranged in opposite pairs. Large flower with maroon-red petals hangs like a pendant and emits a foul smell. It blooms at night and falls in the morning. Fruit is long and dangling, like a gourd. **HABITAT** Introduced to India from Africa. Widely cultivated in India. Prefers moist conditions with well-drained soil. **USES** Suitable for use as a shade tree, and commonly used as an avenue tree.

Indian Cork Tree

■ *Millingtonia hortensis*
Tree. Height: up to 25m

DESCRIPTION Tall, deciduous tree with a feathery crown. Compound leaves are complex with twice or thrice pinnate leaves. Fragrant flowers bloom at night and drop in the morning. They are produced in August–September, and are white, long-tubular shaped and grow in droopy clusters. Fruits are long, flat pods. **HABITAT** Naturalized in parts of central India and often planted in parks. **USES** Used in ornamental woodwork and carvings. Bark is used as a substitute for true cork. Often planted along roadsides. However, it is not suited for use as an avenue tree since it is shallow rooted and its branches are brittle.

Yellow Trumpet Tree

■ *Tabebuia argentea*

Tree. Height: up to 8m

DESCRIPTION Small, deciduous tree, often with a crooked trunk. Compound leaves have a silvery sheen, and have narrow leaflets arranged palmately. Showy, trumpet-shaped yellow flowers appear in clusters. **HABITAT** Native to South America. Cultivated throughout India. Growes taller in moist climates than in dry regions, and is drought tolerant. **USES** Attractive flowering tree, mainly planted as an ornamental in parks and gardens. Crooked trunk forms interesting shape.

Pink Trumpet Tree

■ *Tabebuia impetiginosa*
Tree. Height: up to 7m

DESCRIPTION Attractive flowering tree with light deciduous crown. Leaves are compound, and arranged palmately. Flowers are showy in winter when the tree is nearly leafless. Trumpet-shaped flowers are pink-purple with a yellow-centred throat and appear in dense clusters. Fruits are bean-like pods. **HABITAT** From tropical America and the West Indies. It thrives in milder climates, but can also adapt to the harsh extreme temperatures of Delhi. Can tolerate different soil conditions, ranging from clayey and loamy, to sandy. Drought tolerant once established. **USES** Ornamental tree with showy leafless blooms, so good for parks and gardens.

Roheda ■ *Tecomella undulata*

Tree. Height: 5–8m

DESCRIPTION Flowering tree of arid regions, also known as Desert Teak. It is deciduous, and the compound leaves are narrow, wavy and grow in opposite pairs. Attractive when in blossom, its large flowers appear in March. They are trumpet shaped and vary in colour from yellow to orange. Bean-shaped pods contain winged seeds. **HABITAT** Native to India. Occurs in semi-arid to arid conditions of North-west India. Grows in difficult climates and conditions. **USES** Important timber species used in furniture making. Bark has medicinal value, and is used in many herbal and Ayurvedic formulations. Shallow, lateral-spreading roots help to bind soil and prevent soil erosion.

African Tulip Tree

■ *Spathodea campanulata*
Tree. Height: 12–20m

DESCRIPTION Ornamental tree with evergreen foliage. Sheds leaves in dry climates. Dark green leaves are compound, and arranged in opposite pairs. Bell-shaped, large flowers are orange-red and appear with new leaves in March–April. Fruits are sharply tapering pods that contain papery seeds. **HABITAT** Native to tropical regions of Africa, and widely cultivated in India. Prefers moist conditions. **USES** Ornamental tree planted in city parks and avenues. Shallow, lateral-spreading roots help in containing soil erosion.

Yellow Trumpet Bush ■ *Tecoma stans*

Shrub. Height: up to 3m

DESCRIPTION Large shrub with dense green foliage. Leaves are compound, and leaflets are lance shaped with serrated edges in opposite pairing. Trumpet-shaped flowers are deep yellow, flowering throughout the year. **HABITAT** Hardy shrub native to tropical parts of America, and naturalized in India. **USES** Suited for planting along roadsides, and also pruned and maintained as a hedge. Widely grown in parks and along avenues.

Chamrod ■ *Ehretia laevis*

Tree. Height: 5–9m

DESCRIPTION Small tree of delicate proportions – its light-coloured bark and knotted trunk are distinctive features. Leaves are simple, oval with a sharp tip and arranged alternately. Small white flowers appear in clusters in spring. Blossoms open in the evening and fall by dawn, covering the ground beneath the tree in a carpet of fine white flowers. Fruits are small berries that turn yellow when ripe in loose, drooping clusters.
HABITAT Occurs in dry, deciduous forests of India. Can survive harsh conditions and has a root system that can grow straight out of rocks. It is drought and frost tolerant.
USES Root and bark are known to be medicinal. Leaves are used as fodder for cattle. Fruit is edible, although not tasty; it is eaten in times of famine or scarcity.

Freshwater Mangrove

▪ *Barringtonia racemosa*
Tree. Height: 10–15m

DESCRIPTION Coastal tree, with a straight-growing trunk and a dense, rounded crown. Large, leathery leaves are simple, oval in shape with a blunt-pointed tip. Flowers are fragrant, similar to powder puffs, comprising numerous white stamens that are pink at the ends. They are night blooming. Fruits are curiously shaped, almost conical but angular. **HABITAT** Tree of moist tropical regions, it grows near riverbanks and swamps, associated with mangroves in less saline conditions. **USES** Quick growing, it is used as an avenue tree along city roads.

Cannonball Tree

Couroupita guianensis
Tree. Height: 20–30m

DESCRIPTION Large tree with curious round fruits hanging on the trunk, which is its distinguishing feature. Leaves are simple, large and oval in shape, with pointed ends. They are shed a few times a year. Flowers are attractive and highly fragrant, and appear in April; the scarlet petals are large with whitish-yellow centres. Woody fruits are spherical and resemble cannonballs. They have an unpleasant smell when ripe and open. **HABITAT** Native to Guyana in South America, and cultivated in moist tropical parts of India. **USES** Used as an exotic ornamental tree. It has religious significance to Hindus and is often planted near temples. Hard shells of the fruits are sometimes used as kitchen utensils.

Ber ▪ *Zyzyphus mauritiana*

Tree. Height: 5–8m

DESCRIPTION Deciduous tree with a low, spreading crown. Dark green leaves are simple, with silvery underleaves covered in fine hair. Flowers are minute and whitish-green, appearing in clusters a couple of times a year. Oval fruits are fleshy and edible, and green ripening to yellow-orange in colour. **HABITAT** Native to India and most parts of Asia, and cultivated widely for its fruits. Hardy plant that thrives in dry climates and can tolerate waterlogging. **USES** Fruits are rich in vitamins, and are eaten fresh, dried or preserved. Local communities use wood as firewood and leaves as fodder. Leaves, fruits and bark are used in medicinal remedies to cure stomach ailments. When closely planted, this spiny tree can be used as a living fence.

Hemp ■ *Cannabis sativa*

Shrub. Height: 2–3m

DESCRIPTION Erect shrub with attractive foliage. Leaves are compound with leaflets having serrated edges, and are arranged palmatley. Flowers are minute and grow in racemes. **HABITAT** Native to Central Asia and parts of the Indian subcontinent, and can be found growing in the wild in wasteland. **USES** Cultivated widely for its medicinal and narcotic uses. Flowers and leaves are used medicinally. Fibre from the stem is used in making ropes, fabric and paper. Hemp oil is used in paints and varnishes.

Canna ■ *Canna indica*

Shrub. Height: 1–2m

DESCRIPTION A common shrub that is often seen along drains and marshes. It is a leafy shrub that grows out from a rhizome. Large leaves resemble banana leaves. Flowers are on a spike and have small, narrow, red to yellow petals. Hybrid cultivars have broader petals, whereas wild species have narrow ones. Fruits are spiky, rounded capsules containing seeds with hard shell covering. **HABITAT** Native to tropical America, and naturalized in many parts of the world. Suited to growing in full sun, in marshy, wet situations. **USES** Attractive bed plant for gardens. It is being successfully used in constructed wetland, where the root system of the plant helps to cleanse the water.

Barna ■ *Crataeva nurvala*

Tree. Height: 10–12m

DESCRIPTION Small tree often with crooked branches and a light crown. Leaves are compound and comprise three leaflets that are oval with acute pointed tips. Tree is deciduous and sheds its leaves in winter. Dramatic blossoms are eye-catching; white flowers turn yellow with time and have long, antennae-like stamens. They are produced in March–May. Fruits are globular, with seeds encased in a pulp. **HABITAT** Occurs naturally in dry, deciduous forests in large parts of India. Prefers moist growing conditions alongside rivers and streams, and can remain stunted in dry regions. **USES** Bark and roots are medicinal, and used as an appetizer and laxative. Leaves and bark are also used as a poultice to treat rheumatism. Wood is utilized to make small utility objects such as combs and matchsticks.

Cashew Nut Tree ■ *Anarcardium occidentale*

Tree. Height: up to 8m

DESCRIPTION Small tree of unruly growth cultivated for its nutritious nuts. It is semi-deciduous and has large leaves. Leaves are oval in shape with rounded tips, and rough in texture. Flowers are small and inconspicuous, white in colour turning red with time. Fruit is composed of two parts, a fleshy stalk on the top holding a kidney-shaped nut protruding below. Both parts are edible. **HABITAT** Native to tropical parts of America and cultivated widely in coastal regions. India is known to be one of the largest producers of cashew nuts. **USES** The fleshy stalk of the fruit is referred to as cashew apple and can be eaten fresh, or fermented to make liquor. The nut is nutritious and much in demand. Kernel covering the nut is toxic and should be eaten after roasting.

Mango ▪ *Mangifera indica*

Tree. Height: 20–30m

DESCRIPTION Evergreen tree with dense, rounded crown, grown for its celebrated fruits. Simple leaves are long with pointed tips shaped like a lance. Emits aromatic smell when crushed. Flowers are minute, greenish-white and arranged in clusters. They appear in January–March. Oval-shaped fruits are green, ripening to yellow-orange. Size and shape of fruits differ according to the variety of mango. **HABITAT** Native to India. Found wild in the forests of central India, the foothills of the Eastern Himalaya and the Western Ghats. Cultivated widely thoughout India. **USES** Wood is considered an inferior-quality timber, and is used to make furniture. Mango is one of the most celebrated fruits in the country, available in many varieties and tastes. Medicinally, the leaves and bark have antioxidant properties. A summer drink made from unripe mangoes is drunk as a coolant. The tree also has a religious significance, and its leaves are used in Hindu rituals. Due to its dense crown, the tree is suited for use as a shade tree.

Amla ■ *Phyllanthus emblica*

Tree. Height: 10–15m

DESCRIPTION Medium-sized, deciduous tree with drooping branches. Simple leaves are small and oblong, and arranged in opposite pairs. Minute flowers are produced in clusters in March–April. Fleshy, rounded fruits with segments grow around a hard nut, in large clusters. **HABITAT** Found in deciduous forests of India and cultivated throughout the subcontinent. **USES** Cultivated for the fruits, which are a rich source of vitamins and minerals. Most parts of the tree, including the flowers, fruits, roots and bark, are medicinal. The fruit contains rejuvenating properties and is used in *Chyavanprash;* it is also one of the ingredients of the Ayurvedic preparation *Triphala.*

Lime ■ *Citrus aurantifolia*

Tree. Height: up to 5m

DESCRIPTION Small fruit tree with low, spiny branches. Dark green glossy leaves are strongly aromatic. White flowers are fragrant, purplish in bud and pollinated naturally by insects and bees. Showy edible ball-like to oval fruit ranges from green to yellow in colour. **HABITAT** Native to India, southern China and South-east Asia. Widespread across India which is one of the world's largest lemon and lime producing countries. It is drought tolerant. **USES** The fruit is very popular in all manner of food for its sour juice. The rind also finds use in cooking, especially baking. The fruit has very high vitamin C content. Largely used in food but also used for cleaning household items and grease from dishes. In earlier times it was used to make citric acid.

Pomelo ■ *Citrus maxima*

Tree. Height: 5–15m

DESCRIPTION Small, evergreen fruit tree with low, spreading branches. Leaves are compound, oblong in shape and arranged alternately. White flowers are highly fragrant. Fruit is large and rounded like a ball. Pulp is sour-sweet, and has white or sometimes pink or red flesh, and a thick, spongy rind pith. **HABITAT** Native to south and South-east Asia, growing from the Sub-Himalayan area to South India. Cold tolerant and does well in saline conditions in light to medium, well-drained soil. **USES** Often grown in gardens for its unusually large, delicious edible fruits. Like all citrus fruits this too has a high vitamin C content. Fragrant flowers are used in perfume making.

Curry Leaf

Murraya koenigii
Tree. Height: up to 6m

DESCRIPTION Small tree with compound leaves that have small, aromatic leaflets arranged in opposite pairs. Flowers are small, produced in clusters and faintly scented. **HABITAT** Native to tropical parts of Asia. **USES** Often maintained as a large shrub in home gardens. Most Indian homes have one either in a pot on a balcony or in a kitchen garden. Widely used for culinary purposes in most Asian cuisines. Leaves, roots and bark are also used in tonic to treat stomach ailments.

Orange Jasmine

Murraya paniculata
Shrub. Height: 3–4m

DESCRIPTION Bushy, evergreen shrub that can grow into a small tree. Dense glossy green foliage is made up of compound leaves; leaflets are oval in shape with acute pointed tips. Flowers are produced in clusters and are highly scented. Its petals are white and fleshy like those of the citrus family. Fruits are small and berry like. **HABITAT** Native to tropical parts of Asia, and widely cultivated throughout India. Grows in full sun and also in partial shade. **USES** Mostly grown as a hedge, it is clipped when young to promote a bushy trunk. It is sometimes left untrained to grow into a beautiful tree.

Indian Quince ■ *Aegle marmelos*

Tree. Height: 10–15m

DESCRIPTION Medium-sized tree, with a deciduous crown. Leaves are compound, comprising three leaflets, and oval shaped with an acute pointed tip. Flowers are white and fragrant, similar to others in the citrus family. They appear in April–May. The woody fruits are round and large, and contain numerous seeds in yellow orange pulp. They take many months to ripen on the tree and hence the large hanging fruits are its characteristic feature. **HABITAT** Native to India, distributed in rainforests. **USES** The ripe orange pulp of the fruit is edible and eaten fresh or made into a summer drink that is considered nutritional and a coolant. It also possesses medicinal properties as an astringent, anti-inflammatory and healer of wounds. The pulp is used as a binder in lime mortar recipes used in construction works. Tree is of significance to the Hindu faith and associated with Shiva temples, where the leaves are used as votive offerings.

Coffee ▪ *Coffea travancorensis*

Shrub. Height: 3–3.5m

DESCRIPTION Large, evergreen shrub with dark green, glossy leaves. Flowers are hermaphroditic and sweet scented; corolla is white, tubular and normally has five lobes. Fruits are red or purple, edible cherries used for making coffee. **HABITAT** Native to tropical Africa – southern Ethiopia, South Sudan and possibly east tropical Africa (Kenya, Mt Marsabit). In India grows in humid evergreen forests of the Western Ghats around Mysore, Coorg and Kerala. **USES** Coffee is made from the roasted beans of the seeds of the coffee plant. It is one of the world's favourite drinks and a very valuable cash crop.

Gardenia

▪ *Gardenia jasminoides*

Shrub. Height: 3m

DESCRIPTION Ornamental shrub with dense green foliage and a dense rounded form. Evergreen, glossy leaves are attractive and have prominent veins, arranged in opposite pairs. Large white flowers are fragrant, borne single at the tips of branches in April–May. **HABITAT** Part of a large genus of 200 species, spread around tropical Asia and southern Africa. *Gardenia jasminoides* is native to parts of South-east Asia, and popularly planted in gardens throughout India. Prefers acidic soil conditions and locations in partial shade, although it does well in full sun as well. **USES** Widely cultivated in Indian gardens as an ornamental.

Fire Bush ■ *Hamelia patens*

Shrub. Height: 1–2m

DESCRIPTION Large, spreading bush that can grow to 5m tall. Fine leaves are mostly elliptical in shape with a short, pointed tip. Leaf colour is medium green in late spring, but changes to purplish reddish-bronze in the North Indian winter. Scarlet flowers are tubular in shape and bloom virtually throughout the year in clumps. **HABITAT** Native to South America and Florida, USA. Hardy plant that is salt tolerant and can adapt to most soil conditions. **USES** Used across India in gardens as an ornamental plant that attracts butterflies and hummingbirds. Leaf is used to treat skin conditions, fever and headaches in its native region.

White Ixora ■ *Ixora parviflora*

Shrub. Height: up to 3m

DESCRIPTION Tall, evergreen shrub that is part of a genus of 400 species, ranging from shrubs to small trees. Leaves are elliptical and glossy green. White flowers appear in dense clusters that have a sweet fragrance, in March–April. Popular varieties of Ixora have flowers of colours ranging from yellow, orange, and pink, to red. **HABITAT** Distributed from the western peninsula to Orissa, all the way to Burma in the east. Thrives in warm climate and is susceptible to frost. **USES** Cultivated widely in gardens across India for its scented flowers.

Egyptian Star Cluster

■ *Pentas lanceolata*
Shrub. Height: 0.6m

DESCRIPTION Low shrub with spreading form and attractive flower clusters. Leaves are bright green, lance shaped and covered with fine hair. They are arranged in opposite pairs. Small, star-shaped flowers are tubular and grow in tight, half-round clusters of varying colours, ranging from white and pink, to red and mauve. **HABITAT** Native to tropical parts of Arabia and Africa. It is grown widely in Indian gardens. Prefers a mild climate with a location in semi-shade and is susceptible to frost. In hot climate like that of Delhi, it is grown as a winter seasonal plant. Thrives in moist conditions with well-drained soil. **USES** Planted as an ornamental groundcover in gardens or as a pot plant.

Rondeletia ■ *Rondeletia odorata*

Shrub. Height: up to 2m

DESCRIPTION Evergreen, woody shrub. Dark green leaves are simple, elliptical in shape with a pointed tip, and arranged in opposite pairs. Flowers appear in summer months, in tight clusters of small orange tubes with prominent yellow rim in the centre. They are faintly scented, rich in nectar and attract birds. **HABITAT** Part of a large genus from tropical America and West India, of which only a few species are cultivated. *Rondoletia odorata* is native to Panama and Cuba. Thrives in the Indian summer and prefers locations with full sun but also tolerates partial shade. **USE** It is hardy and adaptable to various climate and soil conditions and suited to Indian gardens.

Kadamb ■ *Anthocephalus cadamba*

Tree. Height: up to 35m

DESCRIPTION Large tree with an upright trunk and horizontal branching. It has a distinct form. It is deciduous, and its large leaves are simple, oval in shape and have pointed tips. The veins are prominent in the leaves. Flowers are fragrant, in the form of a globe, comprising numerous tiny buds. The flowers drop off leaving a bare globe, a fruit that is eaten by many birds and other animals. **HABITAT** An indigenous tree of India, it is distributed in the Sub-Himalayan tracts of North-east India. It is also popular in city parks and avenues throughout India. **USES** The ripe orange pulp of the fruit is edible and eaten fresh or made into a summer drink that is considered nutritional and a coolant. It also possesses medicinal properties as an astringent, anti-inflammatory and healer of wounds. The tree is of significance to the Hindu faith and associated with Shiva temples where the leaves are used as votive offerings. An essential oil from the flowers is known to be used in making perfumes.

Kaim ■ *Mitragyna parvifolia*

Tree. Height: up to 20m

DESCRIPTION Deciduous tree with a rounded crown. Simple leaves vary in size and shape, and are arranged in opposite pairs. Minute, fragrant flowers appear in May–June. They are clustered spherically on long stalks to form a globe. Fruits are tiny pods clustered around a sphere. Leaves and flowers resemble those of *Anthocephalus cadamba*, commonly known as the Kadamb. **HABITAT** Native to India, occurring in deciduous forests. Favours moist conditions. **USES** Wood is medium hard and durable, and used in light interior construction and carpentry work. Leaves are used as fodder. Bark and root are utilized in local medicinal remedies.

Arjun ■ *Terminalia arjuna*

Tree. Height: up to 24m

DESCRIPTION Evergreen tree often seen planted along roads. Leaves are simple, oblong and have pointed tips; they are arranged in opposite pairs. Minute flowers are inconspicuous, appearing on spikes. Fruits are of an interesting star shape, prominently drooping in clusters. **HABITAT** Native to India. Grows naturally along water bodies and streams, and also cultivated. **USES** Leaves used as food for silkworms. Bark and gum have medicinal applications due to their astringent properties. Brown dye from bark is used for tanning. Grown in parks and as avenue tree.

Baheda ■ *Terminalia bellirica*

Tree. Height: 25–30m

DESCRIPTION Deciduous tree of large proportions, with a spreading crown and often supported on a buttressed trunk. Simple, large, oval-shaped leaves are arranged at the ends of branches. Minute flowers are greenish-white clustered along a spike, and appear in October–November. Fleshy, oval-shaped fruits are eaten by animals when ripe. **HABITAT** Distributed widely in South-east Asia, it occurs in deciduous forests and favours dry conditions. **USES** Leaves are valuable for fodder. Fruits are astringent, and pungent. One of the three ingredients of an Ayurvedic preparation called triphala. Yellow dye is extracted from the fruits, used in tanning leather and as a fabric dye. Oil from the kernel is used in making soaps.

Indian Almond

■ *Terminalia catappa*
Tree. Height: 15–25m

DESCRIPTION Large tree of distinct form, with branches growing horizontally in defined tiers. Large leaves are oval with rounded tips, and clustered at the tips of branches. They are semi-deciduous, and turn yellow and red before shedding. Whitish-green flowers are small, and clustered along curved spikes. Fruit is large, almond shaped and green. **HABITAT** Coastal tree of moist tropical climates, growing in sandy, loamy soil. Naturalized in India. **USES** Fast-growing tree often planted along avenues and in parks as a shade tree. Most parts of the tree are rich in tannin and produce a yellow and black dye. Wood is used in general carpentry works. Leaves are used to feed silkworms. Kernel around the seed is edible once toasted. It yields an oil used as a substitute for almond oil.

Crepe Ginger ■ *Costus speciosus*

Shrub. Height: up to 3m

DESCRIPTION Attractive foliage shrub with large, erect leaves. It is a tuberous plant growing up from rhizomes. Long leaves are acutely pointed with silky hair underneath. Showy white flowers with yellow centres and dark maroon bases have crinkly petals, hence the name Crepe Ginger. Variegated and dwarf cultivars are also to be found.
HABITAT Native to India and parts of South-east Asia, these are moisture-loving plants.
USES Attractive garden plant used as an ornamental. Rhizomes are medicinal and have astringent properties. They can also be cooked and eaten, and are rich in starch.

Henna ■ *Lawsonia inermis*

Shrub. Height: 4m

DESCRIPTION Tall, woody shrub with scant foliage, growing with upright branches. Leaves are tiny, lance shaped and arranged in opposite pairs. Minute flowers are faintly fragrant. **HABITAT** Native to Arabia and Persia, and cultivated in dry regions of India such as Rajasthan, Punjab and Gujarat. **USES** Leaves produce an orange dye, commonly called mehendi or henna. It is used for colouring the hair, producing decorative patterns on the palms and feet, and sometimes for dyeing fabrics. Leaves and flowers yield an essential oil used in perfumes. Leaves are known to have astringent properties and are used in Ayurveda and Unani medicines.

Crepe Myrtle

■ *Lagerstroemia indica*

Shrub. Height: 4–5m

DESCRIPTION Attractive flowering shrub with a light, deciduous crown, can grow fast to a small tree. Small leaves are light green and elliptical in shape. They turn red and yellow before they are shed. Flowers are small, appearing in clusters at the ends of branches. Flower petals are crinkled and occur in variety of colours – white, pink, red and mauve. **HABITAT** Not native to India as the scientific name suggests, but from China and Japan. Hardy plant that adapts to varying climate conditions and is widely cultivated in India. **USES** Mainly planted as an ornamental and widespread in Indian gardens. Bark and roots are medicinal and have stimulant and astringent properties.

Queen's Crape Myrtle ■ *Lagerstroemia speciosa*

Tree. Height: 10–15m

DESCRIPTION Medium-sized deciduous tree that is celebrated for its profuse flowering. It has simple, large leaves with prominent veins, arranged in opposite pairs. Leaves turn red before they are shed. Attractive, mauve-purple flowers with crinkled petals grow in clusters, and cover the crown of the tree. They are produced in April–June. Spherical fruits grown in clusters turn woody when ripe. **HABITAT** Native to tropical Asian countries. Occurs in the forests of the Western Ghats, and north-east and southern India. Cultivated widely throughout the subcontinent. Favours moist, damp climate and grows along riversides and water bodies. **USES** Wood is strong and durable, and used for door and window frames, flooring and making furniture. It is known to be durable underwater. Most parts of the tree have medicinal value. Root and bark are used in remedies to cure stomach ailments. Poultice of leaves is used to reduce fever. Tree is grown in parks and gardens as an ornamental.

Pomegranate *Punica granatum*

Shrub. Height: 3–5m

DESCRIPTION Large, deciduous shrub with unruly branches, which can grow to be a small tree. Leaves are small and oblong, and arranged in opposite pairs. Scarlet flowers can vary towards orange, and have crinkled petals; they are usually borne solitarily or in twos and threes. Fruits are large and rounded, with a crown-like tip at the bottom. **HABITAT** Originally from the Mediterranean region, and cultivated throughout India. **USES** Grown for its delicious fruits and also as an ornamental garden plant. Fruits are a rich source of vitamins and iron. Fruit, rind and roots are medicinal. Used in Ayurvedic medicine as an appetiser, digestant and blood purifier. Also used to treat heart conditions. Fruit and rind are rich in tannin and flowers yield a red dye.

Custard Apple

■ *Annona squamosa*
Tree. Height: 4–6m

DESCRIPTION Small, semi-deciduous fruit tree with a light crown. Leaves are elliptical, pointed towards the end but not sharp. They release an aroma when crushed. Flowers are yellow-green with fleshy petals. Fruits are sweet, delicious and fragrant. They have a rounded form with a scaly, undulating outer surface, which is divided into many segments. **HABITAT** Fruit tree from South and Central America, widely cultivated in India for its fruits. **USES** Fruits are best eaten fresh, and provide a very popular flavour to *Sitaphal* ice cream.

Hari Champa ■ *Artabotrys odoratissimus*

Shrub. Height: up to 3m

DESCRIPTION Medium-sized scrambling shrub to large, woody climber, depending on habitat. Narrow, elliptical leaves are shaped like a lance and glossy green in colour. Greenish-yellow flowers have leathery petals and a very strong fragrance. Flowering occurs from early summer to the rainy period. Smooth, ovoid fruits are 3–4cm long. **HABITAT** Native to India and tropical Asia. **USES** Oils used in perfume industry. Fruit and bark used to treat ailments including fever, diarrhoea, dysentery, bruises, cuts, pain, sprains and inflammation in traditional medicine systems.

Ashok ■ *Polyalthia longifolia*

Tree. Height: 10–15m

DESCRIPTION Tall, straight-growing tree with a narrow, tapering crown of evergreen leaves. The dark green leaves are glossy, long and narrow; they are shaped like a lance with wavy edges. Flowers are greenish-yellow and inconspicuous. The oval-shaped fruits appear in clusters and turn purple when ripe. **HABITAT** Ashok is indigenous to parts of South India and Sri Lanka. It is widely cultivated in Indian gardens and parks. **USES** It is most often planted in straight rows along pathways and roads. Due to its slender and dense form, it is grown as a screen to cut off noise and sightlines along plot boundaries. The leaves are strung together with flowers as decorations during Hindu ceremonies.

Elephant Apple Tree

Dillenia indica

Tree. Height: 10–15m

DESCRIPTION Robust, evergreen tree with a low, spreading crown. Dark green leaves are large and simple, with prominent veins and serrated edges. They are the distinguishing feature of the tree. Flowers are white with yellow centres, faintly fragrant, and attract bees and other insects. Large, round fruits are edible and known to be favourites among elephants in forests. **HABITAT** Native to the evergreen forests of the east and south of the Indian Peninsula, and large parts of South-east Asia. Grows into a very large tree in its natural habitat. **USES** Fruits are eaten and preserved for use in pickles. Also used medicinally as a tonic and laxative. Bark and leaves are astringent. Leaves are used as feed for silkworms. Planted as an ornamental shade tree in city parks and gardens.

Sal ■ *Shorea robusta*

Tree. Height: 20–30m

DESCRIPTION Tall forest tree with a straight-growing trunk. Large leaves are simple and deciduous, and oval in shape with pointed tips. Produces dramatic clusters of fragrant, cream-white flowers. Oval fruits with wings are a true characteristic of the dipterocarp family. **HABITAT** Native to India. Found in dry to moist deciduous forests, as well as in evergreen forests. Favours moist, sandy soil with good drainage. **USES** Important timber tree used in construction and – popularly – for railway sleepers. Heartwood is durable and resistant to termites. Leaves are used as fodder, and oil yielded by seeds is used for cooking. Broad leaves are used to make plates.

Poison Arrow Plant *Acokanthera spectabilis*

Shrub. Height: 2–3m

DESCRIPTION Evergreen shrub with a wide spreading crown. Dark green leaves are stiff and leathery, arranged in opposite pairs. Small white flowers are borne in clusters at the tips of branches and are intensely fragrant. All parts of the plant are poisonous except for the purple berries, which are relished by birds. **HABITAT** Hardy, versatile, drought-tolerant plant that is sun loving but also tolerates shade. **USES** Favoured in Indian gardens for its heady scent but not common. Traditionally used as poison for the tips of arrows.

Desert Rose *Adenium obesum*

Shrub. Height: 1–3m

DESCRIPTION Ornamental shrub with showy, succulent leaves arranged in spirals at branch ends. Flowers are tubular, varying from pink to red with a white blush. They resemble the flowers of plumerias. Evergreen and drought resistant. Known to live for hundreds of years. **HABITAT** Native to the Middle East and tropical and subtropical Africa, and naturalized in many parts of the world. Prefers sunny locations. **USES** Sap from the roots and stem is poisonous and is used as an arrow poison in Africa. In India, planted in gardens as a decorative plant.

Saptaparni ■ *Alstonia scholaris*

Tree. Height: 25–30m

DESCRIPTION Tall, straight-growing tree with a dense, evenly formed crown. Leaves are glossy green, arranged in whorls around the tips of the stem. Flowers are small and in clusters on slender branchlets. They have a strong, cardamom-like fragrance and appear in October–December. Fruits are slender, bean-like folicles. **HABITAT** Native to Indian subcontinent, South-east Asia, and parts of China (Yunnan) and Australia (Queensland). **USES** Owing to its hardy nature and quick growth, Saptaparni has become a popular avenue tree in India. This tree is used in traditional systems of medicine. The plant is used as a drug to help restore the digestive system. It is also useful in relieving fevers and skin ailments, and has been used as an astringent in Ayurvedic medical science for centuries.

Easter Lily Vine ■ *Beaumontia grandiflora*

Vine. Height: 7–9m

DESCRIPTION Thick, woody vine with broad, deep green leaves that are evergreen, oval and pointed at the tips with prominent veins. Large, bell-shaped white flowers are fragrant, and appear in January–March. Twining and very showy when in full bloom. **HABITAT** Native to the Indian subcontinent, China and South-east Asia. Prefers full sun, but tolerates partial shade. Hardy and prefers a moist, hot climate. **USES** Cultivated as an ornamental vine.

Milkweed ■ *Calotropis gigantea*

Shrub. Heigh: up to 4m

DESCRIPTION Erect shrub with large, succulent leaves that are oval and have pointed ends. Greyish-green in colour. White flowers appear in loose clusters, on and off throughout the year. C. *procera* varies from C. *gigantea* in the colour of its flowers, which have a purple tinge. Swollen, dry pod fruits have numerous seeds covered with silky fibres. **HABITAT** Found across India and from Iran to South China, growing in wasteland and in cultivation. **USES** Host plant of butterflies. In India associated with the god Shiva, and flowers are used as votive offerings. Valued for its medicinal properties, and the latex, flowers and leaves are used in Ayurvedic remedies. Known be effective in treatments relating to digestion and skin conditions.

Sea Mango ■ *Cerbera manghas*

Tree. Height: 8–10m

DESCRIPTION Small, evergreen coastal tree with a spreading crown, associated with mangroves. Leaves are simple and long, with pointed tips – the milky sap is poisonous. Showy white flowers are fragrant and have pink-yellow centres. Oval-shaped fruits are poisonous. **HABITAT** Occurs along the coast and backwaters of the Indian Peninsula. **USES** Wood is utilized to make charcoal. Bark is used medicinally for its laxative properties. Poisonous leaves and fruits are sometimes used to poison and catch fish.

Champa ■ *Plumeria* spp.

Tree. Height: 3–9m

DESCRIPTION There are many species of *Plumeria*, with differences in the shapes of the leaves or the colour of the flowers, and variations in size from shrubs to medium-sized trees. Common to all are the large, oblong leaves that are simple and clustered towards the branch ends. *Plumeria acuminata* has leaves with pointed ends, whereas the leaves of *P. obtusa* are rounded. Attractive flowers are fragrant and occur in many colours, ranging from white, to pink to red. Flowers contain no nectar to attract bees and birds; they depend on moths for pollination. Fruits are long, oval-shaped pods, rarely seen on the tree. Soft branches and leaves contain a milky sap. **HABITAT** Native to tropical America (central America and Mexico). Widely cultivated throughout India and South-east Asia. **USES** Mainly planted as an ornamental in gardens and parks. Flowers are used as votive offerings and planted near temples; also referred to as Temple Tree.

Plumeria acuminata

Plumeria pudica

Plumeria obtusa

Plumeria rubra

Plumeria acuminata

Oleander ■ *Nerium oleander*

Shrub. Height: 2–3m

DESCRIPTION Large, evergreen shrub that is common along Indian roads. Narrow, lance-shaped leaves. Flowers are profuse in summer and faintly fragrant. Colours vary from white, cream, yellow and pink to red. All parts of the plant are highly toxic, and cows and goats do not eat them. **HABITAT** Hardy plant that can survive poor soil conditions and is drought tolerant. Widely cultivated in most parts of India. *Nerium* is distributed in warm climate regions of northern Africa to south-west Asia. **USES** Suitable for roadside planting, and also grown in parks and gardens. Oleandrin present in most part of the plant is used medicinally to stimulate the heart and also as a diuretic.

Chandni

■ *Tabernaemontana coronaria*

Shrub. Height: up to 3m

DESCRIPTION Large shrub with attractive, spreading crown that can grow to a small tree. Evergreen leaves are glossy green and attractive. White flowers appear like pinwheels in loose clusters several times in a year. Some varieties are fragrant. **HABITAT** Found across India except in the mountains and distributed widely in South-east Asia. Sun-loving plant that also does well in partial shade. **USES** Suited for growing as a hedge plant, and if not pruned can grow to be a small tree with profuse flowering. The dwarf cultivar with smaller leaves makes a good border plant in a garden. Wood is used in making incense. Chewing the roots is known to relieve toothaches.

Copperleaf ■ *Acalypha wilkesiana*

Shrub. Height: 2–3m

DESCRIPTION Evergreen shrub with foliage in varying shades of green and copper. Leaves are heart shaped and serrated. Male and female flowers appear on the same plant and can be differentiated by their length; male flowers grow on long, drooping spikes, female flowers on short spikes. **HABITAT** The origins of this shrub are in tropical islands of the Pacific Ocean, but it is grown widely around the world. Prefers a semi-shaded location in the garden. **USES** Popular garden plant in India. Cultivated for its brightly coloured foliage, and mainly clipped and maintained as a hedge.

Euphorbia ■ *Euphorbia* spp.

Shrub. Height: up to 3m

DESCRIPTION Euphorbias are a large genus of 2,000 species of shrub, succulent and herbaceous plant. Succulent species are spiny and resemble cacti. A common characteristic is the flowers that have no petals, only stigma and stamen. The stems contain a milk sap that is toxic. **HABITAT** Most of the succulent species are endemic to Africa and some are from India. They are capable of growing in arid rocky areas.

Euphorbia royleana

Crown of Thorns ■ *Euphorbia bojeri*

Shrub. Height: up to 1.5m

DESCRIPTION Spiny shrub in which the thorns are arranged around the spine in a spiral. Leaves are oblong and evergreen, though they are shed in dry conditions. Flowers are greenish-white, minute and arranged on a cluster of rounded bracts that are scarlet to pink. **HABITAT** Native to Madagascar and cultivated widely across India. Hardy shrub tolerant of drought and saline soil conditions. **USES** Often used as a living fence in boundary walls, and suited for growing both as a hedge and as an ornamental.

Poinsettia ■ *Euphorbia pulcherrima*

Shrub. Height: up to 3m

DESCRIPTION Ornamental shrub with multiple stems growing upright. In humid, moist climates, it can grow to be a small tree. Leaves are long and lance shaped, and arranged at the ends of branches. Flowers are minute and creamy-white, produced in clusters offset by large, colourful bracts. Bracts are scarlet and showy, and also occur in different colours – yellow, and pink in cold weather from autumn to early spring. **HABITAT** Native to Mexico and Central America, and cultivated throughout India. **USES** Hardy and quick growing, and planted as an ornamental.

White Leaf Poinsettia ■ *Euphorbia leucocephala*

Shrub. Height: 2–3m

DESCRIPTION Delicate version of Poinsettia (see above), with leaves and bracts that are smaller than the red bracts of Poinsettia. Small flowers are clustered at the ends of branches with white bracts, turning the whole shrub to a showy white. **HABITAT** Native to Mexico and Central America, and cultivated in India, though not common. Favours moist, well-drained soil. **USES** Planted as a specimen ornamental plant in gardens.

Jatropha ■ *Jatropha curcas*

Shrub. Height: up to 4m

DESCRIPTION Useful shrub with large, decorative foliage. Leaves have about five shallow lobes, and are deciduous. Flowers are minute and inconspicuous. Fruits are rounded in clusters, ripening to a yellow colour. **HABITAT** Native to tropical parts of America. Widely cultivated in India and can be seen growing in wasteland. **USES** Fruit is used in a variety of ways. Oil from the seed is flammable and burns with soot; it is used as a fuel. Studies are being conducted in its use as a biofuel. It is also used medicinally to treat skin conditions and rheumatism. Fruit is poisonous and is sometimes used to kill fish. Sometimes planted as a hedge.

Coral Plant ■ *Jatropha multifida*

Shrub. Height: 2–3m

DESCRIPTION Attractive shrub with a loose, spreading crown. Leaves are large and decorative, with multiple sharp lobes almost dividing the leaf palmately. Coral-red flowers are small and appear in bunches in the summer months. **HABITAT** Plant of semi-arid regions native to Mexico and Central America. **USES** Grown in Indian gardens as an ornamental. In its region of origin, it is used in a variety of medicinal remedies.

Breadfruit ■ *Artocarpus altilis*

Tree. Height: 15–20m

DESCRIPTION Tropical evergreen fruit tree. Its defining feature is the dense crown of large, deep green leaves, which are deeply lobed and cut into five or seven serrations. Male and female flowers appear on the same tree and are pollinated by bats. Fruits are large, similar to those of Jackfruit (see p. 74) but round in shape. Numerous fruits of many flowers unite to form a single large fruit, a characteristic of syncarps such as the Breadfruit tree. **HABITAT** Widely cultivated in South-east Asia and the Pacific region. Grows best in deep, alluvial soil, and also tolerates coastal sand. **USES** Fruits are sweet and aromatic, and eaten fresh when ripe; unripe fruits are cooked. Also dried and pounded to make flour, which is used to make bread, hence the tree's common name. Inner bark fibre is used to make ropes. Wood is strong and used in construction as beams and posts.

Jackfruit ■ *Artocarpus heterophyllus*

Tree. Height: 10–15m

DESCRIPTION Evergreen tree with dark green, lustrous foliage, and a short, straight trunk topped by a large, dense crown. Large leaves are simple, oval and pointed at the ends, and have a leathery texture. Fruits are very large, heavy and oblong in shape, with a spiny, textured skin. They contain many seeds covered in a sweet pulp that is fragrant when ripe. **HABITAT** Distributed in the Western Ghats, preferring moist growing conditions. Cultivated for its fruits in most parts of India. **USES** Unripe fruits are cooked as a vegetable, and ripe ones are eaten fresh. Sweet and aromatic ripened fruits are also cooked to make jackfruit kheer – a unique recipe in Kerala. The seeds can be roasted and eaten, and are a rich source of starch. Wood is used in carpentry works.

Fig Tree *Ficus carica*

Tree. Height: up to 9m

DESCRIPTION Small tree with low-spreading, twisted branches. The broad leaves are simple and lobed and form 3–5 segments. Figs are shaped like pears, and the green fruits turn deep purple when ripe. Most fruits of the fig family are edible and eaten by birds and other animals and this species is the one preferred by humans. **HABITAT** Native to the Mediterranean region and cultivated in India for its fruit. It is most suited to a sunny location with well-drained, sandy soil. Mature and established trees are drought resistant. **USES** The fig is sweet and nutritious and can be eaten fresh, or dried and preserved.

Banyan ■ *Ficus benghalensis*

Tree. Height: 15–20m

DESCRIPTION Large, evergreen tree with a spreading canopy that sometimes covers acres of land. Strangler with aerial roots, it spreads by means of roots that lower to the ground and develop into trunks, thus forming a colony of trunks. Large, oval-shaped leaves are simple and have a leathery texture. Fruits are formed from a cluster of flowers (the inflorescence), with each flower in it producing a fruit; the flowers mature into a single mass. Each fig thus forms an enclosure with numerous fruits within it. This is a curious phenomenon of the fig family, in which the flowers are pollinated by highly specialized wasps that dig inside the fruits to lay eggs. **HABITAT** Grows widely throughout India, and found wild in the Sub-Himalayan forests and western peninsula. **USES** Figs are edible and attract a large number of birds and mammals. Bark, leaves, latex and buds are medicinal, and used in Ayurvedic formulations. Ropes are made from the fibre of the aerial roots. The tree is of deep cultural and religious significance to Indian communities, and it is planted in many public squares and temples.

Peepal ■ *Ficus religiosa*

Tree. Height: 18–20m

DESCRIPTION Deciduous tree with rounded crown. Leaves are simple, shaped like hearts and have acutely pointed tips. New leaves are a flush of red and show prominent veins. Flowers and fruits are typical of the fig family and are borne directly on the branches. **HABITAT** Grows widely throughout India. **USES** Of religious significance to both Hindu and Buddhist religions. A venerated tree best known as the Bodhi Tree under which Gautama Buddha attained enlightenment. A sapling from the original Bodhi Tree in Bodhgaya is known to survive in Sri Lanka – it is claimed that it is more than 2,000 years old.

Rubber Tree ■ *Ficus elastica*

Tree. Height: 20–30m

DESCRIPTION Tree of gigantic proportions with aerial roots that snake down to form buttresses to support the large form. The leaves are large, oval in shape and have a thick, leathery texture. New leaves appear with red sheaths wrapped around them, rendering the tree visually distinct. Figs are similar to others in the family and oval in shape. **HABITAT** Its native habitat is the foothills of the Eastern Himalaya in the north-east of India. The tree is an epiphyte by nature. **USES** A white latex is drawn from the tree to manufacture rubber and therefore it was widely cultivated. This practice has been discontinued as there are now other viable sources of rubber. Today, Rubber Tree is grown for its large, glossy foliage of many variations, as an ornamental and shade tree.

Pilkhan ■ *Ficus virens*

Tree. Height: 20–25m

DESCRIPTION Large tree with a dense, spreading crown. The trunk is short, topped by an overbearing crown of lush foliage. A network of aerial roots wraps around the trunk of the tree. Leaves are simple, elliptical in shape and taper to pointed tips. They are shed at the end of winter and new leaves appear in spring in various shades of copper, maturing to a fresh green. Figs are small in size and light in colour, often referred to as the white fig. **HABITAT** Found in the Sub-Himalayan range of North-west India and parts of Central India. **USES** Fast-growing shade tree, often planted as an avenue tree along roads and in parks.

Mulberry ■ *Morus alba*

Tree. Height: 6–9m

DESCRIPTION Medium-sized, fast-growing, deciduous tree with a dense, spreading crown in summer. Simple leaves vary in size and shape, and tree is leafless in winter. Flowers are minute and inconspicuous. Fruits are long and fleshy, and comprise minute berries arranged on a spike. They are greenish-white in colour, but there are varieties that produce reddish-purple berries. **HABITAT** Native to large parts of Asia and cultivated throughout India for its fruits. **USES** Fruits are eaten fresh and can be preserved in jams. Leaves are used to feed silkworms, and this is an important tree in sericulture. Bark, roots and fruits have medicinal value. Wood is used in making play equipment such as hockey sticks and racquets. Makes an ideal shade tree due to its dense crown.

False Myrhh *Commiphora wightii*

Shrub. Height: 2–3m

DESCRIPTION Straggly, woody shrub that grows to a small tree. Deciduous, with a distinctive silvery bark that comes off in flakes. Aromatic leaves are shiny green above and greyish below. Gum is exuded from the bark. Flowers are small and brown-red. **HABITAT** Indigenous to India and Sri Lanka, stretching east to China. Tree of semi-arid regions of Rajasthan, Gujarat and Karnataka that does well in rocky and sandy soil conditions. **USES** Resin extracted from incisions in the trunk and branches is called Guggul, and is used in the Ayurvedic medicine system, incense and perfume making. Medicinally it has astringent, antiseptic and anti-inflammatory properties.

Alexandrian Laurel

Calophyllum innophyllum

Tree. Height: up to 30m

DESCRIPTION Coastal tree with a rounded crown. Large, leathery leaves are evergreen, oval in shape and have rounded ends. Attractive white flowers with a yellow centre appear in clusters in May–August. Greenish oval fruit is a hard nut containing a single seed, eaten by fruit bats. **HABITAT** Occurs along the coast on sandy beaches of southern India, extending to parts of South-east Asia and Australia. **USES** Planted as an ornamental in parks and gardens, and commonly seen as an avenue tree in cities. Wood is strong and durable, and used for building boats and as railway sleepers. Dye is obtained from the bark, which contains tannin. Oil extracted from the seeds is used for treating skin conditions and rheumatism.

Kokum ■ *Garcinia indica*

Tree. Height: up to 9m

DESCRIPTION Evergreen fruit tree. Simple leaves are dark green and glossy, elliptical in shape with pointed ends, and arranged in opposite pairs. Greenish-white flowers are small and inconspicuous, and produced in November–February. Fruits are spherical, ripening to a dark purple colour. Seeds are encased in a pulp. **HABITAT** Native to India, and found in evergreen forests of the Western Ghats and the Konkan region. Cultivated widely in South India for its fruit. **USES** Wood is suited to cabinet making. Rind of the fruits is used in cooking and also to make a juice known as Kokum, which is known to be a coolant. An edible fat, called Kokum butter, is derived from the seeds. Leaves and fruits have astringent and digestive properties, and are used in medicinal remedies.

Tamal *Garcinia xanthochymus*
Tree. Height: 5–9m

DESCRIPTION Evergreen tree with a rounded, spreading crown. Leaves are narrow and shaped like lances. Flowers are small and white, and produce small, round fruits with pointed tips. They turn deep yellow on ripening and the seeds are encased in an acidic pulp. **HABITAT** This tree's natural habitat is in northern India and the Western Himalaya. **USES** Fruits are edible and used in preserves like jams and pickles. Of religious significance to the Hindus and often planted near Krishna temples.

Iron Wood ■ *Messua ferrea*

Tree. Height: 20–25m

DESCRIPTION Slender, evergreen tree with beautiful foliage. Leaves are simple, shaped like lances and arranged in opposite pairs. New leaves are tender red and drooping. Mature leaves have a silver-grey underleaf. Showy, fragrant flowers with white petals and a yellow centre appear in February–June, depending on the part of the country. Oval, rounded fruits have pointed tips. **HABITAT** Native to India and parts of South-east Asia. Distributed in the tropical rainforest of east and south India, and in the Western Ghats. **USES** Hard and durable wood is referred to as ironwood. It is considered to be twice as hard as teak, and is often used for railway sleepers and in heavy construction. Leaves and flowers have antibacterial properties and are used in medicinal remedies. Seed oil and kernel are used to treat skin conditions. Flowers are used in dyeing and also as votive offerings in Hindu temples. The tree is often planted near temples.

Shell Ginger *Alpinia zerumbet*

Shrub. Height: 1.8m

DESCRIPTION Hardy shrub with long, pointed leaves that are dark green in colour. Variegated version with yellow-striped leaves is also available. Flowers are pearly white, borne in drooping clusters. Individual flowers are similar to small seashells and are therefore referred to as 'shell ginger'. **HABITAT** Native to east Asia. Very adaptable, growing in varying climate conditions. Can grow in light shade to full sun, and prefers moist, rich, fertile organic soil. **USES** Planted as an ornamental foliage shrub in gardens.

Turmeric *Curcuma longa*

Shrub. Height: up to 1m

DESCRIPTION Leafy shrub that grows perennially. Long leaves are oblong, and dark green on top and pale below. Flowers are yellow-white and borne on spikes. Fruits are small and ovoid. **HABITAT** Tropical plant that needs some moisture along with moderately high temperatures. The plant has been grown and used in India for at least 2,500 years, and has been cited in ancient Sanskrit texts. **USES** Bright yellow powder obtained from rhizomes is used in food, dyes, cosmetics and medicines. It is an essential part of Indian cuisine, and is used to treat skin conditions and wounds, as well as digestive ailments. Turmeric also has cultural and spiritual significance, associated with fertility and prosperity, and is used in Indian weddings, when it is applied to a bride's face and body as part of the ritual purification before a wedding. Turmeric powder is also sprinkled on sacred images.

Indian Crocus ■ *Kaempferia rotunda*

Shrub. Height: 0.3m

DESCRIPTION Herbal plant that remains dormant in winter. It is a deciduous shrub. White-purple flowers appear when the shrub is leafless, directly from the ground – hence the plant's other name Bhumi-champa, or Flower of the Earth. **HABITAT** Found in East and North-east India, and many parts of South-east Asia. **USES** Rhizome has many medicinal properties and is used to treat stomach disorders. Young leaves and rhizomes are used for culinary purpose.

Ginger ■ *Zingiber officinale*

Shrub. Height: up to 1m

DESCRIPTION Bushy shrub with multiple shoots that arise from rhizomes. Stems comprise series of leaves tightly wrapped around each other. Leaves are deciduous, narrow and up to 7cm long. Flowers are rare on cultivated plants, and are pale yellow with a purple lip and yellow spots. **HABITAT** Possibly native to India, this shrub grows in humid, partly shaded areas across tropical and subtropical zones of India. **USES** Widely grown in India as a commercial crop, and used in food, drinks, cosmetics and medicines. Rhizome is aromatic and is the source of ginger, which has been used as a food flavouring for centuries. Medicinally, it is used to treat colds and coughs, and is considered an appetiser.

Azalea ▪ *Azalea indica*

Shrub. Height: 2–6m

DESCRIPTION Evergreen or deciduous shrub, depending on cultivar. Leaves are small and pointed. Flowers are bell or funnel shaped, and often fragrant. Flowers profusely in spring, with flower colours ranging from white, pink and pale yellow to crimson red. Leaves and flowers, including the nectar, are toxic. **HABITAT** Native to the Eastern Himalaya and grows in other regions such as Europe and North America. Azaleas favour growth under or near trees, and are shade tolerant. They are fussy plants and prefer an environment that is neither too hot nor too cold. In India they are very popular in the north-east region, where they are a favourite with the nursery trade. They grow best in soil that is humus rich, well drained and acidic.

Drumstick ■ *Moringa oleifera*

Tree. Height: 10–15m

DESCRIPTION Common dedicuous tree with uneven form and light crown, today feted as a source of superfood. Compound leaves are complex, comprising leaflets that are further divided into leaflets in opposite pairs. Small white flowers appear in clusters 2–3 times a year. Fruits are long pods, like beans, and are commonly known as drumsticks. **HABITAT** Native to India. Can tolerate a wide range of climates and soil types, and cultivated in most of India. **USES** Drumsticks are an essential ingredient of South Indian cooking. New leaves and flowers are also cooked as vegetables. Considered to be one of the superfoods because it is rich in minerals and vitamins. Bark and leaves are medicinal. Leaves are ground and applied to skin to reduce pain and swelling. Seeds yield an edible oil, and bark is used in tanning.

Hydrangea ■ *Hydrangea macrophylla*

Shrub. Height: up to 1.5m

DESCRIPTION Bushy shrub with dense, rounded form. Leaves are simple and elliptic in shape, with pointed tips. Small flowers are gathered in a showy ball, with petals of varying colours – mauve, pink and white, sometimes with a blue tinge. Colour of flowers depends on soil conditions – acidic soil produces blue flowers, while alkaline soil produce pink ones. **HABITAT** Native to Japan, but popular around the world as a garden shrub. It has a somewhat limited geographical reach in India, where it is a cold-climate plant that is often used in the hills and mountains as a garden plant. Grows vigorously and is easy to maintain. Thives in full sun as well as semi-shade, and prefers acidic soil with moist conditions. **USES** In Japan the leaves are used to make a liquor. Young leaves can be cooked and eaten. Tea made from dried leaves is used in certain Buddhist ceremonies.

Magnolia ■ *Magnolia soulengiana*

Tree. Height: 6–8m

DESCRIPTION Small, deciduous tree with spreading crown of shiny green leaves. They are simple, oval shaped and arranged alternately. Showy flowers are fragrant, with pink petals that are white inside. New variants occur in white, pink, rose, purple, magenta and burgundy colours. A hybrid plant initially bred by the Frenchman Etienne Soulange-Bodin by crossing *Magnolia denudata* with M. *liliiflora*. **HABITAT** Indian hill stations and temperate areas. Good specimen trees can be found in Srinagar Valley, and in Kalimpong in the Eastern Himalaya. Grows in full sun to part shade in moist, well-drained, acidic soils. **USES** Planted in gardens as an ornamental flowering tree.

Southern Magnolia ■ *Magnolia grandiflora*

Tree. Height: 20–25m

DESCRIPTION Ornamental evergreen tree with large, glossy green leaves, which are simple, with a leathery texture. Large flowers are highly fragrant with creamy-white petals arranged like a rosette. Cones are conspicuous, with red seeds. Magnolias are a large genus of over 100 species, and are considered to be ancient plants. They are slow growing. **HABITAT** Native to south-east America, and cultivated in India as an ornamental. Thrives in moist, cool climate. **USES** Used as an ornamental in parks and gardens. Essential oil from bark is used to treat malaria and rheumatism.

Golden Champak *Michelia champaca*

Tree. Height: 25–30m

DESCRIPTION Large-leaved, evergreen tree. Simple leaves are acutely pointed and arranged in spirals. Fragrant flowers appear in April–May, and their colours range from creamy-white to orange. Fruits are oval shaped and clustered along a spike. **HABITAT** Native to India, and distributed along the foothills of the north-east Himalaya and Western Ghats. Popularly cultivated in the south for its flowers. **USES** Fine-grained wood is used in furniture making. Bark, leaves and flowers are used in medicinal remedies. Flowers are used as votive offerings and are often worn in the hair as an adornment. Essential oil from flowers is utilized in perfume making. The tree is planted near temples and as an ornamental in parks.

Neem ▪ *Azadirachta indica*

Tree. Height: 10–15m

DESCRIPTION Medium-sized tree with a nearly evergreen crown; one of the most common trees planted in India. The feathery leaves are compound in nature, and the leaflets are lance shaped with serrated edges and arranged in near-opposite pairs. Flowers are small and white in loose clusters and are faintly fragrant. Oblong fruits are berry-like, turn yellow when ripe and contain one seed. **HABITAT** Widely cultivated and naturalized in most parts of India; it occurs wild in the dry forest of the Deccan Peninsula, Myanmar. **USES** Wood is very durable and used in general carpentry works. Leaves contain a compound called azadirachtin that repels insects. Dried leaves are kept in cupboards to keep away moths. Neem oil from seeds is used in soaps. The bark of roots and stem, leaves, and oil are medicinal. Neem twigs are chewed and used as toothbrushes. The seedcake after oil extraction is used as an excellent manure.

Persian Lilac ■ *Melia azedarach*

Tree. Height: up to 15m

DESCRIPTION Medium-sized tree with a light deciduous crown. Leaves are compound with leaflets arranged in opposite pairing. They are similar to Neem leaves but have sharply serrated edges. Minute flowers appear in March–April, and are faintly fragrant. White petals are tinged with a hint of mauve, and have a purple centre. Green berries turn yellow when ripe, and droop in bunches – a characteristic of this tree. **HABITAT** Native to a large part of South-east Asia, including India. Can adapt to various soil and climatic conditions. **USES** Wood is used to manufacture agricultural tools and tool handles. Leaves are used as fodder. Persian Lilac is a quick-growing tree, useful for planting as a shade tree in gardens and parks. It is, however, short lived.

Baobab ■ *Adansonia digitata*

Tree. Height: up to 20m

DESCRIPTION An unusual tree with a large bulbous trunk supporting a network of thin branches reaching upwards. It is a deciduous tree and the branches remain bare at most times. New leaves appear in spring; they are compound and arranged palmately. Flowers are large and white with a central column of stamens; they hang facing downwards on long stalks. They bloom at night and fall to the ground by day. The large fruits are woody when ripe and contain angular seeds in an acidic pulp. **HABITAT** The species is exotic to India, planted as a specimen tree and is native to Sub-Saharan parts of Africa. Known to be one of the longest living trees, with some individual trees claimed to be over 2,000 years old. **USES** Mostly planted in city parks as a specimen tree. The swollen trunk retains water in its tissues and over time becomes hollow. The hollows are known to be large enough to be inhabited by tribal people. Parts of the tree are also medicinal. The pulp of the fruit is good for the heart. The leaves are ground to a paste and applied to reduce swelling. In Ayurvedic medicine, the bark is considered to be astringent, an appetiser and a cooling agent.

Silk Cotton Tree ■ *Bombax ceiba*

Tree. Height: 20–30m

DESCRIPTION Large tree with a spiky trunk supported by buttressed roots. Branches are arranged in tiers, with a light crown of leaves. Compound leaves are palmate – a common characteristic of the mallow family. Flowers are large and showy, with thick, waxy petals whose colours range from yellow to scarlet. They bloom in February–March and are rich in nectar, attracting a variety of birds and bees. Fruits are angular, woody pods containing seeds attached to a silky floss. **HABITAT** Native to tropical Asia, and widely distributed in India in both dry and moist deciduous forests. **USES** Planted in city parks and gardens as an ornamental tree. Silky cotton is used as stuffing and as insulating material. Flowers are eaten as a vegetable by certain communities. Gum exuded from the bark, called *mocarasa*, is used in Ayurvedic medicine.

White Silk Cotton

■ *Ceiba pentandra*
Tree. Height: 20–25m

DESCRIPTION Large, deciduous tree with distinctive, spiky green trunk, developing large buttressed roots with age. Compound leaves comprising 5–8 leaflets are arranged palmately, and are shed in winter. Creamy-yellow flowers are small and inconspicuous, blooming at night. Fruits are long pods, tapered at both ends. They open when ripe to release seeds that are attached to a silky floss. **HABITAT** Tropical tree from South America and South Africa. Planted widely in India. **USES** Silky floss is used to stuff bedding and cushions. It is water resistant, so is also used in lifebuoys and jackets. Oil extracted from the seed is known as *kepok* seed oil. Tree is planted in city parks and along avenues.

Floss Silk Tree

Chorisia speciosa
Tree. Height: 10–12m

DESCRIPTION Attractive medium-sized tree, distinguished by a greenish spiky trunk. Light crown is deciduous, with palmate compound leaves. Showy flowers appear when the tree is leafless in winter. They have pink petals with yellow centres. White-flowering variety also occurs, but is not common. Fruits are woody pods that open to release seeds with a silky floss. **HABITAT** Tree of South America, which thrives in dry conditions and is drought tolerant. **USES** Wood makes good-quality timber and is used for light interior works. Gum yielded by the tree is used as an adhesive. Often planted as an avenue tree.

Kanak Champa ■ *Pterospermum acerifolium*

Tree. Height: 15–25m

DESCRIPTION Tall tree with dense crown of large leaves that are deciduous. Rounded leaves are simple with wavy edges, and covered with silvery fine hair. Large, longish flowers are creamy-white and fragrant. They bloom at night and fall during the day. Their characteristics are typical of flowers that are pollinated by bats. Large, woody pods open up into neat segments to release papery thin seeds. **HABITAT** Occurs in the foothills of the Himalaya. Favours moist situations like banks of rivers and water bodies, where it rarely sheds its leaves. **USES** The moderately hard wood is used in interior construction works. Flowers, bark and leaves have medicinal properties. Large, rounded leaves are used to make plates. Due to its attractive form and foliage, and fragrant blossom, suited for use as an ornamental.

Wild Almond Tree

■ *Sterculia foetida*

Tree. Height: 15–20m

DESCRIPTION Tall tree with attractive foliage. Leaves are palmately compound and clustered towards branch ends. Small flowers appear around March and have a foetid smell, reflected in the scientific name. Large, woody pod is reddish and arranged in clumps, and contains seeds that are edible. **HABITAT** Indigenous to the west and south of India. **USES** Seed can be roasted and eaten, similarly to almonds. Leaves and bark have medicinal properties.

Phalsa ▪ *Grewia asiatica*

Shrub. Height: up to 4m

DESCRIPTION Straggling shrub that can grow to be a small tree. A dwarf variety is popular in cultivation. Leaves are simple and deciduous, broad and with a coarse texture. Small yellow flowers are clustered in groups of 2–5. Fruits are small, round berries, ripening to deep purple or red. **HABITAT** Native to tropical and subtropical India and Pakistan. Does well in habitats with distinct summer and winter seasons. Hardy plant that grows in a wide range of soils and is reasonably drought tolerant. **USES** Very popular for its fruits in India and Pakistan, this plant is much cherished despite its somewhat short fruiting period of about three weeks in May. Ayurveda texts mention many uses for the fruits, including their properties for cooling the body, and curing the heart and blood disease, and as an aphrodisiac.

Hibiscus ▪ *Hibiscus rosa-sinensis*

Shrub. Height: up to 4m

DESCRIPTION Evergreen shrub with large, dark green and shiny, toothed leaves that are arranged alternately. Large flowers are single and bright red, and have a long central tube with stamens and pistils at the tip. Hundreds of varieties are available, with flower colours ranging from scarlet to orange, to yellow to white. **HABITAT** Grows well in most parts of India. Deep, moderately fertile, well-drained, slightly acidic soil is ideal for the plant. **USES** Highly ornamental garden plant. Petals are dried and used to brew a tea. Leaves and petals are used in many home beauty remedies. Used in Ayurvedic medicine to treat diarrhoea and colitis, and to promote hair growth.

Indian Privet

■ *Clerodendrum inerme*
Shrub. Height: 1–2m

DESCRIPTION Dense, evergreen shrub with fine leaves, suited to be maintained as a hedge. Dark green leaves are elliptical with pointed ends, and have a shiny upper surface. They are simple leaves, arranged in opposite pairs or whorls of 3 and 4. Small white flowers are trumpet shaped with prominent pink stamens. Fruits are pear-shaped drupes. **HABITAT** Hardy plant, common across India and distributed widely in South-east Asia. Adapts well to most soil conditions and prefers a spot in full sun. **USES** Popular for hedges in Indian gardens. It is quick growing and is one of the preferred plants to shape into topiary.

Tulsi ■ *Occimum tenuiflorum*

Shrub. Height: 0.5–1m

DESCRIPTION Compact, dense shrub with a rounded form. Leaves are strongly scented. Most common variety is green leaved, but there is a purple-leaved variety known as *Krishna Tulsi*. Flowers are small and purplish, and grow on a spike. **HABITAT** Native to tropical and subtropical India, and normally occurs as a cultivated plant. **USES** Considered a sacred plant by Hindus and grown in the courtyards of most homes and temples. Among its many uses are medicinal ones – it is used for treating a wide range of ailments, from coughs and colds, to stress. In Ayurvedic medicine considered to be an 'elixir of life', and believed to promote longevity. Dried leaves are used for making tea. They are also mixed with food grain during storage to keep insects out.

Bottlebrush ■ *Callistemon lanceolatus*

Tree. Height: 8–12m

DESCRIPTION Evergreen tree of irregular form and with drooping branches. Its deeply fissured trunk and narrow, lance-shaped leaves are its distinguishing features. Dark green leaves have a leathery texture, and release an aromatic smell when crushed. Scarlet flowers are minute, made up of long stamens and arranged in a dense spiral along a spike. **HABITAT** Native to Australia, and cultivated widely throughout India. Moisture-loving plant, growing best along banks of rivers and streams. **USES** An essential oil is derived from the leaves.

River Red Gum ■ *Eucalyptus camaldulensis*

Tree. Height: 15–20m

DESCRIPTION One of the most common tree species in the Indian landscape, this eucalyptus grows tall and has a tapering crown. In its native habitat it can grow to a height of 40m or more. Leaves are long and narrow, and shaped like lances. They contain an aromatic oil and release their fragrance when crushed. A common characteristic of the flowers is that the bud is covered with a cap that comes off when in bloom to reveal a fluff of white stamens. **HABITAT** Native to Australia, from where many eucalyptus species have been introduced to India. Once a popular tree for afforestation due to its quick growth and ability to survive in poor, barren soil conditions. **USES** Wood is straight and long, and is often used for posts. It is also used as pulp in paper making. An essential oil is derived from the leaves.

Jamun ■ *Syzygium cumini*

Tree. Height: 15–20m

DESCRIPTION Tall, evergreen, fast-growing tree with dense foliage. Glossy green leaves are simple and are shed in dry climates. They are oblong with pointed tips, and are arranged in opposite pairs. Whitish-green flowers appear in clusters in March–May. Fleshy berries are deep purple when ripe and delicious. **HABITAT** Native to India and parts of South-east Asia. Thrives in moist situations and does not survive arid situations. **USES** Wood is used for making oars and boats, and also for agricultural tools. Tree is mostly cultivated for its fruits. Leaves and bark are used in treating diabetes, and fruits for treatment of colic. Planted along avenues and in parks as a shade tree.

Yesterday Today Tomorrow *Brunsfelsia pauciflora*

Shrub. Height: 1–2m

DESCRIPTION Bushy shrub with glossy green leaves. The elliptical leaves are pointed at the tips and arranged alternately. The fragrant flowers appear in spring and early summer and have a peculiar habit of changing colours. At first the mauve and white flowers turn purple, then they fade to a shade of white in a day, thus earning the title Yesterday Today Tomorrow. **HABITAT** Introduced to Indian gardens from Brazil. Suited to a sunny location, it also tolerates partial shade. **USES** Often grown in gardens for its colour and fragrance.

Angel Trumpet *Brugmansia suaveolens*

Shrub. Height: 3–5m

DESCRIPTION Evergreen, woody shrub with multiple branches, which can grow to a small tree. Large, oval leaves are pointed at the ends. Remarkably showy and sweetly fragrant, trumpet-shaped flowers hang down in beautiful clumps. They are usually white but may be yellow or pink. **HABITAT** Originally from the rainforests of Brazil, and now popular all over the world as a decorative garden plant. Grows in areas with warm temperatures, heavy rainfall and high humidity, and is happy in full sun. **USES** Planted as an ornamental shrub in gardens. Every part of the plant is poisonous, with the seeds and leaves being particularly toxic.

Day Blooming Jasmine

■ *Cestrum diurnum*

Shrub. Height: up to 2m

DESCRIPTION Dense shrub with multiple upright trunks. Evergreen, with glossy leaves that are simple, narrow and arranged alternately. Flowers comprise inflorescences of small, sweet-smelling white blooms that occur throughout warm weather. Fruits are small black berries. **HABITAT** Subtropical shrub requiring average to moist, light, sandy soil. Naturalized across all regions of India. **USES** Popular garden plant with blossoms attracting many species of butterfly.

Night Blooming Jasmine ■ *Cestrum nocturnum*

Shrub. Height: up to 4m

DESCRIPTION Unruly, evergreen shrub with drooping branches. Small, narrow leaves are simple, and shaped like a lance. Greenish-white flowers have a slender, tubular corolla that opens at night and is intensely fragrant. Blooms in cycles throughout warm weather. Fruit is a small, blackish berry. **HABITAT** Widely cultivated in Indian gardens for its fragrance and was introduced from the Caribbean region of West Indies. **USES** Planted in gardens for its heady fragrance.

Thorn Apple ■ *Datura metel*

Tree. Height: up to 1m

DESCRIPTION Small shrub with dark violet shoots. Leaves are simple and arranged alternately; they are oval shaped with acute pointed tips. Trumpet-like flowers are large and faintly scented. Their colours range from violet, black and white, to red and yellow. Round, dark purple fruits are surrounded by spines. All parts of the plant are poisonous. **HABITAT** Grows wild in warm parts of India, including tropical areas and the temperate Himalaya. Very low-maintainence plant that does well in full sun. **USES** There are numerous religious associations with the plant. Leaves are used as an intoxicant in North India. Most parts of the plant are medicinal and are used in Ayurvedic formulas. Used sparingly due to its narcotic effects.

Yellow Berry Nightshade ■ *Solanum xanthocarpum*

Shrub. Height: 0.3–0.5m

DESCRIPTION Small, prickly bush spreading low on the ground. The leaf and stem are armed with spiky thorns, and thereby commonly known as *Kantakari*. The purple flowers have a yellow centre resembling others of the nightshade family in which five petals are joined together. **HABITAT** Wild-growing plant often seen in dry, open wasteland across India. **USES** The shrub is valued in Ayurveda medicine and known to be effective in the treatment of asthma, coughs and sore throats.

Silverberry ■ *Eleaegnus pyriformis*

Shrub. Height: up to 3.5m

DESCRIPTION Straggly shrub with spiny branches. Leaves are glossy green on top and silvery underneath. They are elliptic in shape with pointed ends. Flowers are small and inconspicuous. Fruits are showy in large clusters, oval shaped and ripen to a bright red colour. The skin of the fruit is mottled with a silvery scale.
HABITAT Found in the foothills of the Eastern Himalaya in North-east India.
USES Mainly planted in home gardens for its delicious fruits. Can be eaten fresh or cooked.

Jasmine ■ *Jasminum* spp.

Shrub. Height: 0.5–3m

DESCRIPTION Plant of subtropical and tropical regions, famed for its sweet-smelling flowers. This is a genus of semi-evergreen to evergreen shrubs and climbers comprising 200 species. The small leaves are glossy green and can be simple or compound. Flowers are star shaped and tubular, but not all species are fragrant. Colours are mainly white, but some are yellow and sometimes pink-red. Jasmine flowers mainly in warm summer months and also makes an appearance in spring or autumn. **HABITAT** Natural distribution of species is mostly in Asia and Africa in warm climates, but it is also grown in colder climate in glasshouses. In India, it is a popular garden plant and is also cultivated commercially for its flowers. Some of the most popular species are *Jasminum sambac, J. multiflorum, J. officinale and J. nitidum*. Thrives in moist but well-drained soil conditions in full sun. **USES** Flowers are used as votive offerings and also strung into garlands to decorate the braids of women. Flowers of *J. odoratissimum* and *J. sambac* are used to flavour tea. Jasmine oil is used in perfumes, incense, cosmetics and soaps. Juice or poultices of leaves of some jasmines are used in various medicinal remedies.

Jasminum nitidum

Jasminum sambac

Yellow Jasmine

■ *Jasminum nudiflora*
Shrub. Height: 1.5–3m

DESCRIPTION Rambling shrub with slender, arching stems that droop to the ground. Dark green leaves arranged in opposite pairs are compound and comprise three leaflets. Bright yellow flowers are unscented and funnel shaped, and appear in April–June. **HABITAT** Common across India in the plains as well as the mountains. **USES** Showy and bold garden plant when in bloom.

Harsingar

■ *Nyctanthes arbor-tristis*
Tree. Height: 4–6m

DESCRIPTION Small tree with drooping branches. Its most attractive feature is its fragrant flowers. Oval leaves are simple, with acute pointed tips, and are arranged in opposite pairs. They have a rough, sandpaper-like texture and serrated edges. Flowers are small and produced in loose clusters in July–October; they have white petals and orange-coloured throats. Fruit is a round, flat pod. **HABITAT** Occurs in the foothills of the Himalaya and in deciduous forests. Widely cultivated in India. **USES** Fragrant flowers are used as votive offerings. Rough-textured leaves are used to polish utensils and wood. Bark and leaves are medicinal, and are used in the Ayurveda, Siddha and Unani systems. Saffron dye is obtained from the flowers, and is used for dyeing fabrics.

Star Fruit Tree ▪ *Averrhoa carambola*

Tree. Height: 6–9m

DESCRIPTION Small tree with dense, attractive foliage. Leaves are compound, with leaflets arranged in opposite pairs. Flowers are minute, and pink to purple in colour. Curiously angular fruits have a cross section shaped like a star. They appear a couple of times in a year, in March–April and November–December. **HABITAT** Origins of the tree are not certain, possibly Indonesia and Malaysia. Cultivated in India but not common. **USES** Sour, sweet fruits can be eaten fresh, juiced or preserved. Due to their acidic quality, they are used to polish brass and other metal objects.

Screwpine *Pandanus odoratissimus*

Shrub. Height: 1–5m

DESCRIPTION Large foliage shrub that, with time, grows a tall stem like that of a tree. Mature plants have numerous prop roots. The bluish-green leaves are long, narrow and sharp like a sword, and can be over 1m long. They are deciduous. Flowers bloom in summer, and male and female flowers occur on separate plants. Only the male ones are fragrant and bloom on a spike. **HABITAT** Found in the wild in South India, and the Andaman and Lakshadweep Islands. Grows from sea level to fairly high altitudes, but prefers moist locations close to seashores and lagoons. It is a hardy plant that is drought and wind tolerant. **USES** Flower is distilled to extract an essence called Kewra, used in food flavouring and drinks. Leaves are used to weave a mat used to thatch roofs. Fragrant flower is sometimes wrapped in cloth and kept in cupboards to perfume clothes.

Pride of Burma ■ *Amherstia nobilis*

Tree. Height: up to 10m

DESCRIPTION Semi-evergreen tree with a spreading crown of attractive foliage. Leaves are compound, comprising long, oval-shaped leaflets in opposite pairs. Tree is most attractive when in bloom, when large clusters of red flowers hang on long stalks. Red petals have highlights of yellow. The fruit is a crimson-coloured, flat pod. **HABITAT** A tree of Burmese origins, this species is not known to occur freely in the wild. **USES** Mostly cultivated as an ornamental flowering tree in hot, humid regions of India such as Kolkata and Mumbai.

Red Orchid Bush ■ *Bauhinia galpinii*

Shrub. Height: up to 3m

DESCRIPTION Sprawling shrub with bushy crown. Leaves are lobed like those of other *Bauhinia* species, and small. Attractive flowers have pink-red petals and resemble flowers of a gulmohar tree. Flowers in summer months. **HABITAT** Originates in South Africa, and in India cultivated in gardens as an ornamental shrub. Favours mild climate conditions. **USES** Ornamental plant in gardens and also a host for butterflies.

Yellow Orchid Tree ■ *Bauhinia tomentosa*

Tree. Height: up to 4m

DESCRIPTION Light shrub growing into a small tree. Attractive leaves are divided into two lobes and are light green in colour. In December–March produces bell-shaped flowers that are bright yellow with a deep maroon-coloured centre. Fruits appear in January–June and are pea-like, slender, velvety and light green when young, turning pale brown with age. Sheds leaves in North Indian cold winters. The genus name *Bauhinia* honours the 16th-century botanist brothers Johann and Caspar Bauhin. They were identical twin brothers, making the genus name very appropriate because the two lobes of the leaves folded together are also identical. **HABITAT** Native to India, Sri Lanka and tropical Africa. **USES** Flowers are rich in pollen and nectar, and attract various butterflies, bees and insect-eating birds. Certain birds and the larvae of some moth and butterfly species feed on the flowers and leaves.

Orchid Tree ■ *Bauhinia purpurea, B. variegata*

Tree. Height: 8–12m

DESCRIPTION Attractive flowering plant that grows to a medium-sized tree. The two species that are the most commonly seen in India are *Bauhinia purpurea* and *B. variegata*, which are deciduous or semi-deciduous trees. Leaves are simple and rounded, deeply lobed and divided into two parts. Leaves of *B. purpurea* have a deeper cleft than those of *B. variegata*. Flowers are showy, ranging from mauve to white with hints of pink. Flower petals of *B. variegata* are broad and showy, and flowers appear in abundance when the tree is nearly leafless. *B. purpurea* flowers are more scant, with narrow petals, and are produced at the same time as the leaves. **HABITAT** Both species are indigenous to India and parts of South-east Asia, occurring in the lower altitudes of the Himalayan Range in dry deciduous forests. **USES** Wood is used to make agricultural tools. Root and bark are used medicinally. Leaves are used as fodder. A natural hybrid, *B. blackeana*, with deep purple flowers, is also commonly planted in Indian gardens as an ornamental.

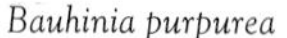

Bauhinia purpurea

Bauhinia variegata

Laburnum ■ *Cassia fistula*

Tree. Height: 10–15m

DESCRIPTION Tree of modest size and form transforming into a highly ornamental one when in full bloom. Compound leaves are deciduous, oval in shape and have pointed ends. Showy flowers are a saturated yellow and are borne in drooping clusters when the tree is leafless, in April–July. Fruits are long, dangling pods, containing many seeds in a pulp. **HABITAT** Indigenous tree of India, occuring in deciduous forests. Hardy and drought resistant. **USES** Bark is used in tanning. Valuable plant medicinally, with the fruit pulp, root, bark and leaves being used as a purgative, laxative and in curing skin conditions. Heartwood is very durable and used to make wheels and agricultural tools. Widely cultivated as an avenue tree and in parks. Planted as an ornamental as well as a shade tree.

Desert Cassia

■ *Cassia biflora*
Shrub. Height: 4–5m

DESCRIPTION Large shrub that can grow to the size of a small tree. It is multi-stemmed with spreading branches floating upwards. Leaves are compound with small, oblong leaflets arranged alternately. Deep yellow flowers are showy, appearing in pairs in summer. **HABITAT** A hardy plant, popular in gardens, it is widely cultivated in India. Its origins are known to be in the tropical parts of America. **USES** Grown as an ornamental for its showy flowers.

Poinciana ■ *Caesalpinia pulcherrima*

Shrub. Height: up to 3m

DESCRIPTION Ornamental shrub with a tall stem and bushy foliage. Leaves are compound, twice pinnate with small, oblong leaflets. Showy flowers are borne on tall, upright spikes and resemble those of the Gulmohr (see p. 121). Petals are bright red tinged with yellow or bright yellow in colour. **HABITAT** Introduced to India from the Caribbean islands of West Indies. **USES** Planted widely in Indian gardens as an ornamentel.

White Gulmohur ■ *Delonix elata*

Tree. Height: 6–10m

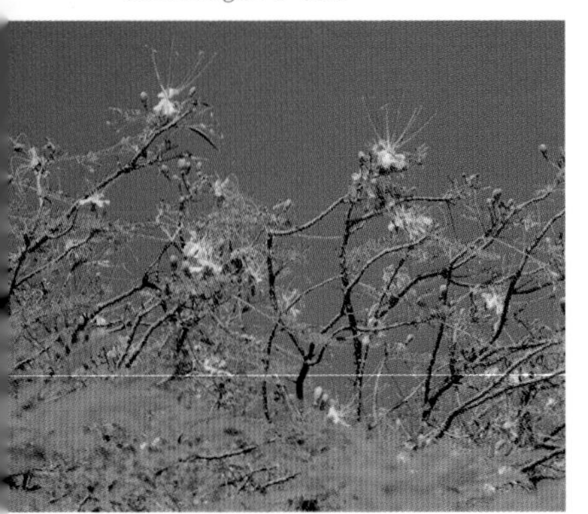

DESCRIPTION Deciduous tree with spreading, rounded crown. Leaves are compound, and twice pinnate with leaflets arranged in opposite pairs. Showy white flowers, which fade to yellow, are formed in clusters in September–March; they have long, pronounced stamens like antennae. Fruit is a long, flat pod, pointed at both ends. **HABITAT** Believed to have been introduced to India from East Africa. Occurs naturally in barren tracts of Gujarat. Prefers a hot, dry climate. **USES** Wood is light coloured, easy to work with and suited for making cabinets. Extract from leaves is used as an anti-inflammatory. Tree is suited to planting for soil conservation and in shelterbelts.

Gulmohur ■ *Delonix regia*

Tree. Height: 10–15m

DESCRIPTION Handsome deciduous tree with light, feathery leaves and a rounded crown. Compound leaves are twice pinnate with leaflets arranged in opposite pairs. Flowers, produced in April–May, are showy, orange to scarlet, and have petals arranged in clusters. Fruits are long, flat pods that droop from the tree. **HABITAT** Although commonly seen, this tree is an exotic, introduced to India from Madagascar. It is cultivated throughout the plains of India as an ornamental, and thrives in dry conditions. **USES** Planted mainly as an ornamental and shade tree in parks and gardens.

Anjan ■ *Hardwickia binata*

Tree. Height: 25–30m

DESCRIPTION Tall tree with a straight trunk and drooping branches. The open crown is light and covered in fine foliage of deciduous leaves. Small leaves are compound and comprise two leaflets joined at the bases. Greenish-white flowers are minute and inconspicuous. Fruits are long, flat pods each containing a single seed, which is carried by the wind. **HABITAT** Native to India, occurring in dry, deciduous forests of central India. Thrives in a dry climate and can grow in difficult soil conditions with rocky ground. **USES** Wood hard and durable, and used in heavy construction for structures such as posts and beams. It makes good fuel and is also used to make charcoal. Fibre from bark is used to make ropes. Leaves are used as cattle fodder.

Sita Ashok ■ *Saraca asoca*

Tree. Height: 7–10m

DESCRIPTION Evergreen tree with a dense, rounded crown. Leaves are compound, and arranged in opposite pairs; young leaves are red and drooping. Orange flowers turn red with time; they are borne in clusters directly from the branches. **HABITAT** Native to India; occurs naturally in moist evergreen forests. **USES** Ornamental tree frequently seen in parks. In certain parts of India wood is used for building and furniture making. All parts of the plant have medicinal properties, and it is widely used in the Ayurveda, Unani and Siddha systems of medicine. The tree has religious significance in both Hindu and Buddhist traditions.

Tamarind ■ *Tamarindus indica*

Tree. Height: 20–25m

DESCRIPTION Large tree, with a straight dark trunk and rounded crown of feather-like leaves. Leaves are compound, with small, oblong leaflets arranged in opposite pairs. They fold up and close at sunset. Flowers are small, on long stalks and droop in loose clusters. The petals are yellow with maroon stripes. Fruits are long and curved pods containing seeds in an acidic pulp. On ripening they turn brown and brittle. **HABITAT** A common tree cultivated throughout India, it was introduced from tropical parts of Africa and has naturalized here. **USES** Pulp is best used after storing for a year when sweetly sour. It is used in Indian cooking to impart a sour flavour, mostly in chutney and sometimes as part of a summer drink. It is considered an appetiser and digestive. In Ayurvedic medicine, most parts of the tree are considered medicinal. The wood is hard and is used for post and beams in buildings.

Black Wattle ■ *Acacia auriculiformis*

Tree. Height: up to 15m

DESCRIPTION Evergreen tree with a straight trunk and open, irregular crown. The leaves are not true leaves but a modification that is designed to reduce water loss from a leaf's surface, referred to as a phyllode. They are narrow and curved with a leathery texture. Flowers are minute, arranged densely on long spikes. They are faintly fragrant. The pods when mature are woody and coil up into curious shapes. **HABITAT** Black Wattle is one of the many acacias native to Australia and also found in New Guinea, and thrives in wetlands near rivers and streams. Being a hardy species that is not browsed by goats and cattle, it was introduced to semi-arid regions of India in an attempt to reforest them. **USES** Due to its fast growth and dense foliage, it is planted as a shade tree on roadsides and in parks. The shallow root system is suited to controlling soil erosion.

Catechu ■ *Acacia catechu*

Tree. Height: up to 9m

DESCRIPTION Small tree with a dark, often crooked trunk. It is a deciduous with compound leaves. Leaves are twice pinnate and comprise numerous small, oblong leaflets. Pale yellow flowers are minute, densely arranged on a long spike and appear in May–June. Fruits are flattish long pods that turn dark brown when mature. **HABITAT** Native Indian tree, distributed widely in the plains and up to an elevation of 1,500m in the Himalaya. **USES** Cultivated for a product called Cutch or Catechu, derived by boiling pieces of the heartwood until they are well reduced and cooled to form a paste. This paste is popularly used to flavour an Indian delicacy called *Paan* and referred to as *Kattha*. Catechu is also used in printing and dyeing techniques. Ayurvedic medicine regards most parts of the tree to be medicinal, and they are used to treat various ailments such as coughs, fevers and wounds.

Babool *Acacia nilotica*

Tree. Height: 7–12m

DESCRIPTION Medium-sized, deciduous tree, often with a dark, crooked trunk and a light floating crown. The branches are spiny and covered with straight, paired thorns. The leaves are compound and twice pinnate, comprising small, oblong leaflets in opposite pairs. Sweet-smelling flowers are a ball of yellow, and appear during early monsoons. The fruit pods are peculiar in shape, long like an interconnected string of beads. **HABITAT** Babool is spread across the plains of India, especially in the arid and semi-arid regions. It can withstand extremely dry conditions as well as flooding. Native to India, Sri Lanka and tropical Africa. **USES** Bark and pod yield tannin, used to tan leather. Wood is durable and termite resistant, and is used to make doors and windows, and also agricultural tools. Leaves and pods are used as fodder for livestock. Gum is used in paper making and calico printing.

Siris ■ *Albizia lebbeck*

Tree. Height: up to 15m

DESCRIPTION Medium-sized tree with a deciduous crown. Most distinct in form when leafless and covered with clusters of loosely rattling seed pods. The compound leaves, like others of the pea family, are twice pinnate, comprising small, oval leaflets. Creamy-white, powder-puff flowers are fragrant and rich in nectar, appearing in April–May. The fruit pods are long and tapered at both ends, and remain on the tree for many months. **HABITAT** Widely distributed tree in the Indian landscape. **USES** Cultivated throughout India as a shade tree due to its hardy, quick-growing nature.

Powder Puff ■ *Calliandra haematocephala*

Shrub. Height: up to 2.5m

DESCRIPTION Upright bush with a spreading crown. Leaves are twice pinnate, arranged alternately. Crimson flowers have long stamens arranged around a spherical head. Different varieties have pink and white flowers. **HABITAT** Native to Mexico and South America, but popular in Indian gardens. Reasonably drought tolerant. **USES** Being a hardy, low-maintenance shrub, often planted along roadsides and in public parks.

Touch Me Not ■ *Mimosa pudica*

Shrub. Height: 0.5m

DESCRIPTION Straggling, low-spreading shrub with a spiny stem. Touch Me Not is so named because a slight touch or movement causes the leaves to fold and withdraw. This is a defensive action taken by the plant to repel herb-eating insects and other animals. The leaves are twice compound and touch sensitive. Flowers are mainly stamens clustered in stalks of a pink-lilac colour and bloom in the rainy season. **HABITAT** Native to tropical America, and found growing weed-like in wasteland and open ground; rarely found in cultivation in India. Known to have been introduced to India in the 16th century, and mentioned in ancient Ayurvedic text for its medicinal use. **USES** All parts of the plant are medicinal. Leaves are ground and used to reduce swelling of glands. Leaf sap can be rubbed on skin to cure sores. It helps arrests bleeding and has wound-healing properties.

Mesquite ■ *Prosopis julifora*

Tree. Height: 10–12m

DESCRIPTION Spiny tree with a dark, crooked trunk and a wide, floating crown. The mimosa-like leaves are compound and twice pinnate with small, oval leaflets arranged in opposite pairs. Flowers are minute, and greenish-yellow in colour. They are densely clustered along a droopy spike. Fruit pods are long and flat and usually curved. **HABITAT** Native of South America, Mesquite was introduced to the arid regions of India in a bid to reforest barren tracts. It is now considered invasive, taking over native vegetations. **USES** Leaves and pods are used as food for cattle. In its native region, a flour is made from the pods after removing the seeds. It is rich in protein and sugar. Wood is collected by local communities for firewood.

Rain Tree ■ *Samanea saman*

Tree. Height: up to 30m

DESCRIPTION Large tree with wide, spreading crown. Compound leaves are small and arranged in opposite pairs. They fold in the evening and during rain, unfolding in the morning to cut off sunlight and provide shade. Pink-white flowers resembling powder puffs appear in clusters. Long pods contain seeds in a pulp that is known to be edible. **HABITAT** Native to parts of South America. Thrives in coastal and tropical parts of India. Tolerates waterlogging. **USES** Wood does not shrink or warp and is used for making craft products. Improves nitrogen content in soil. Mainly planted as a shade tree along avenues and in parks.

Flame of the Forest

■ *Butea monosperma*
Tree. Height: 9–15m

DESCRIPTION Deciduous tree taking interesting forms, often with a crooked trunk. Large, rounded leaves are compound, and comprise three leaflets. Underleaf is covered with fine hair, giving a silvery-grey hue. Flowers are showy, and clustered along a spike. Petals are beak shaped, varying in colour from yellow to deep orange. Fruits are flat pods containing a single seed, as suggested by the species name *monosperma*. **HABITAT** Native to India and parts of South-east Asia. Widely distributed in dry areas and grassland. Can grow in difficult soil conditions and is drought and frost tolerant. **USES** Round leaves are stitched together to make plates. They are also used to wrap *beedi*, a local form of cigarette. Flowers yield a saffron-coloured dye. Tree is suited to hosting lac insects, which produce shellac gum. Wood performs well in water and is sometimes used to line wells. Fibre from inner bark and roots is used to make ropes.

Indian Rosewood ■ *Dalbergia sissoo*

Tree. Height: up to 15m

DESCRIPTION Important timber tree popularly known as *Shisham*. The trunk is often crooked, assuming interesting forms, and crown has delicate foliage. A deciduous tree, its leaves are compound and comprise 3 or 5 leaflets arranged alternately. Flowers appear just after new leaves in March. They are creamy-white and inconspicuous, bearing a faint scent. Fruit pods are flat and narrow and hang on the tree for many months. **HABITAT** Native range is the Sub-Himalayan Tracts along riverbanks and streams. Thrives in moist conditions where it grows to a large tree. **USES** Wood is hard and durable and resistant to termites. It is most loved for its beautiful grain pattern and used in making furniture. Species is hardy and adaptable, and is commonly seen planted along roads as an avenue tree in North India. Oil from the seed is used in remedies to treat skin conditions.

Indian Coral Tree

■ *Erythrina indica*
Tree. Height: 10–15m

DESCRIPTION Straight-growing tree with branches that grow upright. Attractive crown of foliage, which is deciduous. Leaves are compound, with an arrangement of three leaflets that are heart shaped ending on an acute tip. There is a variety with variegated leaves with yellow striations, which is very showy. Flowers are showy scarlet blossoms that grow in a spiky cluster. Fruits are long pods with pointed tips. **HABITAT** Native to the Indian subcontinent, naturally occurring in deciduous forests. Widely cultivated in gardens and parks as an ornamental. **USES** Root system is known to have nitrogen-fixing abilities. Often grown among crops as a support for vines. Red dye is produced from the flowers.

Moulmein Rosewood

■ *Millettia peguensis*
Tree. Height: 10–12m

DESCRIPTION Compact deciduous tree with a droopy form and dense crown. Leaves are compound, comprising seven leaflets arranged in opposite pairs and with a terminal leaflet. Leaflets are oval shaped and pointed at the ends. Flowers resemble those of peas and appear in clusters of lilac in March–April, when the tree is nearly leafless. Fruits are flat and woody pods. **HABITAT** It was introduced to India and its origins are known to be from Myanmar stretching east towards Thailand. Not a common tree in India, it does well in the hot, dry climate of Delhi. **USES** Cultivated as an ornamental in city parks and gardens.

Indian Beech ■ *Pongamia pinnata*

Tree. Height: up to 15m

DESCRIPTION Low spreading tree with drooping branches also commonly known as *Karanj*. At most times, it can be identified by mottled leaves that have been damaged by leaf-mining worms. Leaves are compound, broadly oval in shape and have pointed tips. Flowers appear profusely in clusters after new flush of leaves, and the white petals are tinged with pink and purple. **HABITAT** A versatile tree that can adapt to varying conditions, its origins are in humid environments along seashores and streams. It can tolerate saline and fresh water and in dry inland regions it is also drought resistant. Distributed on the seashores of the Western Ghats, and Andaman and Nicobar Islands. **USES** Often planted along roads as an avenue tree in North India. Pongam oil extracted from the seeds is flammable and is used to light lamps and as an engine lubricant. It is medicinal and applied to treat skin conditions and rheumatism.

Chir Pine ■ *Pinus roxburghii*

Tree. Height: 20–30m

DESCRIPTION Tall, evergreen tree with a tapering crown; it becomes semi-deciduous in a dry climate. Long, needle-like leaves grow in bunches of threes. Male and female flowers are borne separately with females in cones, in February–April. The woody cones take more than two years to mature and release the seeds. **HABITAT** Thrives at an elevation of up to 2,000m in the Himalayan range. Also grows in the plains. **USES** Wood is of inferior timber quality, and is used in carpentry works. Tapped for resin commercially. Suited to being shaped and altered to bonsai plants. One of the fastest growing conifers, which can survive growing on rocky ground. Good for afforestation.

Deodar ■ *Cedrus deodara*

Tree. Height: up to 60m

DESCRIPTION Tall tree with a tapering crown of dense, dark green foliage. Needle-like leaves are arranged alternately in clusters on drooping branches. Male and female flowers are borne separately on different trees. Female cones appear in August and take about a year to ripen. **HABITAT** Tree of high altitudes of the Western Himalaya, at about 1,200–3000m. **USES** Wood is durable, with a characteristic scent, and is used structurally in buildings, for doors and windows, and for railway sleepers. Oil extract is used as an antiseptic. Root system conserves soil and prevents erosion on hillsides. Also planted for its ornamental form and foliage.

Oriental Plane or Chinar

■ *Platanus orientalis*

Tree. Height: up to 30m

DESCRIPTION Tree of large proportions with a dense, spreading crown, known for its longevity. Leaves are sharply lobed, and divided into three, five or seven parts. They are deciduous, turning bright red before they are shed in autumn. Minute flowers are clustered to form a sphere. Fruits are a spiky globe containing a single seed. **HABITAT** Native to south-eastern parts of Europe and south-west Asia. **USES** Wood is used to make furniture and small objects, which are then painted and lacquered in the Kashmiri crafts tradition. Grown in parks and gardens as an ornamental shade tree.

Plumbago ■ *Plumbago capensis*

Shrub. Height: up to 1.5m

DESCRIPTION Low, spreading shrub that can sometimes be trained as a climber. Evergreen with glossy green leaves. In North India the leaves are shed in winter. Flowers bloom profusely in loose clusters and can vary in colour from white to various shades of blue. **HABITAT** Native to Cape area of South Africa, and thrives in warm climates. Does well in light, sandy soil, and tolerates partial shade. It is susceptible to frost and survives best when cut back in winter. **USES** Excellent border and groundcover plant, and a good host for butterflies. Also suited to being trained as a climber on trellises.

Prickly Poppy ■ *Argemone mexicana*

Shrub. Height: 0.1–3m

DESCRIPTION Small, evergreen shrub with angular, sharp leaves that are spiky with a thorny spine. Silvery-grey sheen to green leaves. Flower is a saturated yellow with crinkled petals, similar to a poppy. Fruits are oblong capsules containing brownish-black seeds. **HABITAT** Native to tropical America and naturalized in large parts of India. Found as a wasteland and crop weed. **USES** Most parts of the plant have medicinal value and are used in various remedies in the Ayurveda system of medicine. Latex and leaves are used to cure skin conditions. Latex and powdered root are ingested for stomach disorders. Seed oil, locally known as *Satyanashi* oil, is used for protection from termites.

Camphor ■ *Cinnamomum camphora*

Tree. Height: 10–12m

DESCRIPTION Semi-evergreen tree with a dense, rounded crown. Aromatic leaves are simple, arranged alternately and smell of champhor when crushed. Creamy-white flowers are minute and inconspicuous. Fruits are round berries that turn purple-black when ripe and contain a single seed. **HABITAT** A tree of South-east Asia originating in China and Japan. In India, cultivated in the Nilgiris in the south and the Himalayan foothills in the north. **USES** Camphor is extracted from the stems and root of the plant. It is used in culinary and medicinal preparations, and as a pest deterrent in mothballs. Essential oil from leaves and stem is applied to ease muscular strains and skin inflammation. Smell of camphor is known to have a calming effect.

Peach ■ *Prunus persica*
Tree. Height: 4–8m

DESCRIPTION Small, deciduous tree with drooping branches. Leaves are narrow, long and shaped like lances. Flowers appear when tree is leafless in spring, covering the entire tree in pink blossoms. Stone fruits are fleshy with a yellowish pulp and velvety skin. **HABITAT** Known to have been first cultivated in the Tarim Basin of the Xinjiang region in China, though the name suggests that it came from Persia. Prefers full sun but can tolerate partial shade. **USES** Widely cultivated for its fruits, and also as an ornamental for its dramatic blossoms.

Pear ■ *Pyrus* spp.
Tree. Height: 10–15m

DESCRIPTION Medium-sized fruit tree growing tall with a narrow crown. Oval, glossy green leaves are simple and arranged alternately. White flowers bloom profusely and make beautiful cut-flower arrangements. Fruits ripen on the tree, and are harvested, stored and ripened further to be eaten. **HABITAT** Grows in cool temperate part of India, which is mainly the Sub-Himalayan region. Also grows well on North Indian plains, where there are distinct winters. **USES** Fruits are eaten fresh, baked in dishes and stewed. Also used for making jams and jellies. Wood is excellent for making furniture, woodwind instruments and kitchen utensils, though it is not often used in India. In English gardens the tree is often grown in espalier forms, with the branches being trained to grow flat against a wall or a trellis.

Loquat ■ *Eriobotrya japonica*

Tree. Height: 5–9m

DESCRIPTION Fruit tree with handsome form and evergreen foliage. Large leaves are simple and lance shaped, with prominent veins, and are clustered around branch ends. Creamy-white flowers are small and woolly. They are faintly fragrant and attract honeybees. **HABITAT** Native to south-east China, and cultivated mainly for its fruits in India. **USES** Planted as an ornamental in gardens, and for its sweetly sour fruits.

Apple ■ *Malus domestica*

Tree. Height: 1.8–5m

DESCRIPTION Small, deciduous fruit tree. Simple leaves are arranged alternately and have serrated edges. Flowers are strikingly beautiful, with five petals, and can be white, pink or red. The globular fruit when ripe can have skin colours ranging from red and yellow, to green, pink or russet. **HABITAT** Originating in Central Asia, this tree is grown all over the world. In India the states of Himachal Pradesh, and Jammu and Kashmir, both with temperate climates, are the two main apple-producing areas, from which the fruits are shipped across the country. **USES** Very popular fruit that is eaten raw, and used for making juice, jams and desserts. There are nearly 7,500 cultivars of the cultivated apple.

Tea ■ *Camellia sinensis*

Shrub. Height: up to 15m

DESCRIPTION Evergreen shrub known best for its leaves, which are used to brew tea. Fine, acuminate leaves are fresh green in colour, and are plucked at various stages of maturity depending on the type of tea they are used to produce. In cultivation it is pruned and maintained at a height of 1–2m. **HABITAT** The plant's origins are unclear. Variety called *Camellia sinensis* var. *assamica* is grown in Assam. **USES** Cultivated for its tea leaf. Medicinally, tea is known to be a stimulant. It is used in homeopathic medicine to treat neural conditions. Old tree bushes are culled and the twisted trunk is used as a sculptural base for tables.

Camellia ■ *Camellia japonica*

Shrub. Height: up to 6m

DESCRIPTION Ornamental shrub with attractive, showy blossoms. Leaves are dark green and glossy, and arranged alternately. Flowers are large, in a range of colours from pink to red, and white. Petals are arranged in concentric circles. **HABITAT** Native to China, though it is widely cultivated and a large number of cultivars is available. *Camellia japonica* and C. *reticulata* are two of the important ones. Prefers acidic soils and requires a lot of water; thrives in areas with high rainfall. **USES** Mainly cultivated in gardens as an ornamental.

Tree of Heaven ■ *Ailanthus excelsa*

Tree. Height: up to 20 m

DESCRIPTION Large tree with a distinct branching pattern forming a rounded canopy. Leaves are compound, similar to those of the Neem only larger, due to which the tree is also referred to as *Mahaneem*. The yellowish-white flowers are small and inconspicuous. Fruits are flat pods containing a single seed, hanging in large clusters. **HABITAT** In India, distributed in parts of the Indian peninsula and Central India. A hardy plant that does well in dry, difficult conditions, it is popular in cultivation along roadsides and parks in North India. **USES** Quick-growing, hardy species, it is grown as a shade tree. Light-coloured wood is used in making packing cases.

Bird of Paradise ■ *Strelitzia reginae*

Shrub. Height: up to 1.2m

DESCRIPTION Small, evergreen shrub with attractive foliage. Greyish-green leaves are oval, growing on long stalks and arranged in opposite pairs that fan out in one plane. Flowers are strikingly beautiful and bird-like; the orange sepals and blue-purple petals emerge from a green, boat-shaped bract. Mucus-like substance is produced when in flower. Fruit is a leathery capsule with many small seeds. **HABITAT** Native plant of South Africa popular in tropical gardens around the world. **USES** Ornamental in gardens and a popular cut flower. Flowers are nectar rich and attract bees and sunbirds.

White Bird of Paradise ■ *Strelitzia nicolai*

Shrub. Height: up to 6m

DESCRIPTION Tall shrub similar to a banana plant, with leaves on long stalks spread out like fans in one plane. In structure, resembles **Traveller's Fan Palm** *Ravenala madagascariensis*, but its flowers are a larger version of the Bird of Paradise (see above) and composed of more than one flower. Boat-shaped bracts are purple, and petals are bluish-purple. Seeds are black in colour. **HABITAT** Tropical plant from South Africa, not as commonly seen in India as the Bird of Paradise. **USES** Cultivated for its dramatic form and flower. In its native region, fibre from the stalks is used to make rope. The nectar of the flowers attracts hummingbirds.

Lac Tree or Kusum ■ *Schleichera oleosa*

Tree. Height: 15–20m

DESCRIPTION Large, deciduous tree with a straight, short trunk topped by a large, dense crown. Compound leaves are oval shaped, with pointed tips and arranged in opposite pairs. Tree is conspicuous when new leaves appear in a flush of red in March. Yellowish flowers are minute and appear in clusters; they are followed by berry-like, fleshy fruits. **HABITAT** Mostly found in mixed deciduous forests, the foothills of the Himalaya, and central and South India. **USES** Wood is used to make agricultural tools. Shellac found on the bark is used commercially in varnish, hence the name Lac Tree. Good for providing shade, and often planted in cities along avenues and in parks. Kusum oil extracted from the seeds is used as hair oil and also applied to cure skin conditions.

Himalayan Horsechestnut ■ *Aesculus indica*

Tree. Height: 20–30m

DESCRIPTION Handsome tree of temperate climate with a large, rounded crown. It is deciduous, with compound leaves that are arranged palmately and comprise 5–9 leaflets. Flowers are showy white tinged with pink on tall spikes, and appear in summer around June. Fruits are rounded and spineless, unlike those of other chestnuts and contain a shiny seed. **HABITAT** Indigenous to the North-west Himalaya, the species is found growing on mountains and in valleys at an altitude of 3,000m and is a sight to behold in the Kashmir Valley. It also does well in lower altitudes of Dehradun and Kasauli. **USES** Popularly grown as an ornamental in parks and large gardens for its beautiful foliage and flowers. The wood is used in carvings and making small utility objects. All parts of the plant contain a toxin called aesculin, including the seed. However, the seed is washed, ground and roasted to make a flour. The seed oil is used in traditional medicine. Horses are known to be treated with a local preparation using the fruit, hence the name Horsechestnut.

Soapnut Tree

■ *Sapindus mukorossi*

Tree. Height: 15–20m

DESCRIPTION Deciduous, medium-sized tree with a robust crown. Compound leaves with leaflets arranged in opposite pairs are shed in December. Minute white flowers are inconspicuous. They appear in May–June, and are followed by fleshy green fruits. **HABITAT** Native to China, and cultivated in North India. Thrives in deep, well-drained soil. The species found in South India is *S. emarginatus*. **USES** Fruit contains saponin, a natural detergent, also known as Ritha in India. Pulp is used as a substitute for soap. Good fuel wood suitable for making charcoal. Seeds and fruits have medicinal properties.

Mahua ■ *Madhuca longifolia*

Tree. Height: up to 20m

DESCRIPTION Large, handsome, deciduous tree that is dramatic in new leaf. Large, oval-shaped leaves are simple, with pointed tips, and are arranged at the ends of branches. New leaves are flushed with crimson. Minute, creamy-white flowers, produced in February–April, are fragrant and borne in clusters. They bloom at night and fall by dawn. Fruit is a berry-like green that turns orange-red on ripening. **HABITAT** Native to India, occurring in dry deciduous forests of the Himalayan foothills and central India. **USES** Flowers are consumed in different forms – fresh, dried or fermented as country liquor. They are sought after by birds and mammals, including humans. Often cultivated near villages as a source of food. Leaves and bark are used in medicinal preparations among tribes, and also in the Ayurvedic system. Seed oil is used in soap and candle making, and butter from the seed is edible and used in cooking. Shallow, spreading roots help to prevent soil erosion.

Chikoo ■ *Manilkara zapota*

Tree. Height: 8–15m

DESCRIPTION Medium-sized, evergreen fruit tree with a dense, rounded crown. Deep green leaves are elliptical in shape with pointed ends, and are arranged alternately. White flowers are inconspicuous. Fruits are round and brown in colour, and deliciously sweet and edible. **HABITAT** Tropical tree requiring a hot and humid climate; it can grow up to low altitudes in the Himalayan foothills. Best fruits come from Western India, from the states of Gujarat and Maharashtra, and from Tamil Nadu and Andhra Pradesh in South India. **USES** Popular fruits are available across India, from cities to small villages.

Maulsari ■ *Mimusops elengi*

Tree. Height: 10–15m

DESCRIPTION Straight-trunked, evergreen tree with a dense, rounded crown. Simple leaves are elliptical in shape, with pointed tips and wavy edges. Small, creamy-white flowers bloom at night, and are subtly fragrant. Fruit is an edible, oval green berry that turns deep orange when ripe. **HABITAT** Native to the western peninsula of India. Cultivated in most parts of the country. **USES** Wood is strong and durable, and used in heavy constructions. Oil from seed is used in cooking. Fruits can be preserved and pickled. Fruits, bark and seeds have astringent qualities and are used to treat stomach ailments. The tree has religious significance to Hindus and is associated with Krishna. Often planted near Krishna temples, and also as a shade tree in parks and gardens.

Sandalwood ■ *Santalum album*

Tree. Height: 5–10m

DESCRIPTION Small, slender tree with drooping branches, valued for its scented heartwood and also known as Chandan in India. Smallish leaves are simple, of varying shapes and arranged in opposite pairs. Flowers are minute and inconspicuous, and borne in clusters in February–April. Small, fleshy, spherical fruits, each containing one seed, are eaten by birds. **HABITAT** The origins of Sandalwood are under debate. In India it occurs in dry tropical areas and coastal sand dunes. Young plants need a host plant to provide shade and fix nitrogen levels in the soil. **USES** Heartwood is scented and yields an essential oil. Sandalwood oil is used in cosmetics, perfumes, incense sticks and medicines. Paste of the wood ground in water is applied to the skin as a coolant and antiseptic. Wood is sometimes used as part of the fuel wood in cremations.

Jal ■ *Salvadora oleoides*

Tree. Height: 6–9m

DESCRIPTION Small, semi-evergreen tree forming knotted trunks often bent in shape. Compound leaves are narrow and lance shaped, growing in opposite pairs. Minute flowers appear in March–April, and grow in spikes. Small berries that follow in June are edible. **HABITAT** Native to India. Occurs in arid regions of North-west India. Can grow in rocky dry areas and tolerates saline soil. **USES** Local communities use the wood for fuel, and leaves are a grazing source for cattle. Suitable for growing as windbreaks in desert areas.

Peelu ■ *Salvadora persica*

Tree. Height: 5–7m

DESCRIPTION Evergreen tree of arid regions. Compound leaves of Peelu differ from those of Jal (see above), being broader and more elliptical, with pointed tips. Small, inconspicuous flowers appear in March–April. Small, translucent berries turn pink-red on ripening. **HABITAT** Native to arid regions of North-west India. Also seen on the coasts and seasonal floodplains. **USES** Leaves are a grazing source for cattle and are also eaten in certain communities as vegetables. Fruits are sometimes fermented for liquor. Chewing the twigs is known to prevent dental decay. Suitable for growing as windbreaks in desert areas.

Silver Oak ▪ *Grevillea robusta*

Tree. Height: 15–20m

DESCRIPTION Straight-growing tree with a tapering crown. Feathery leaves are compound, with narrow, pointed leaflets arranged alternately, and silvery underleaves. Orange-yellow flowers are minute, and arranged in showy, spiky clusters. Flowers have no petals but are made up of long, curled-up stigmas. **HABITAT** Native to Australia and cultivated widely in India as an ornamental. **USES** Planted in city parks and along avenues as an exotic. Also grown as a shade tree in tea and coffee plantations. Flowers are a rich source of nectar for bees.

Duranta *Duranta erecta*

Shrub. Height: up to 4m

DESCRIPTION Dense, evergreen shrub with multiple stems; can grow to a small tree. Oval leaves are small, and grow in opposite pairs or in whorls of three. Small flowers are tubular on long, drooping spikes, and are blue to violet in colour. Bunches of yellow berries are showy and poisonous to humans. **HABITAT** Hardy plant that grows across India. Native to Central and South America, and the West Indies. Can adapt to varying soil conditions, and does best in full sun. **USES** Fast-growing shrub suitable for use as a hedge. Flowers attract butterflies and hummingbirds.

Lantana *Lantana camara*

Shrub. Height: 0.5–2m

DESCRIPTION Hardy, common shrub with aromatic leaves. Leaves have finely serrated edges and are covered in hair. Small flowers appear nearly throughout the year, in dense clusters of many colours, ranging from yellow-orange-red to white-pink-red. **HABITAT** Native to tropical America, and has naturalized in many parts of the world, including India. Considered an invasive species and can be found growing in wasteland. It is drought resistant. **USES** Colourful garden shrub that flowers profusely. Attracts many species of butterfly, and is grown in butterfly gardens. Can also be trained as a hedge plant.

Teak ■ *Tectona grandis*

Tree. Height: 20–30m

DESCRIPTION Tall tree with straight trunk, mainly cultivated for its timber. Large leaves are deciduous and simple, oval shaped and arranged in opposite pairs. Flowers are showy white, and minute in size but arranged in frothy clusters. **HABITAT** Native to India and parts of South-east Asia, growing mainly in deciduous tropical forests. Suited to varied habitats, from arid places to moist forests. The growth and height of the tree depends on the region where it is grown. **USES** High-quality timber is valued in construction due to its durability, resistance to pests and beautiful grain. It is mostly used for furniture, doors and windows. Seeds and flowers are used in local medicinal remedies.

Himalayan Poplar ■ *Populus ciliata*

Tree. Height: up to 30m

DESCRIPTION Straight, tall tree with branches growing upright and narrow. Leave are simple, heart shaped and deciduous; they are shed in autumn and the tree remains leafless in winter. Male and female flowers occur on different trees, appearing on long spikes. Seed is dispersed by wind. **HABITAT** Temperate tree of cooler climates, distributed along the Himalaya, most commonly in the Kashmir Valley. Prefers moist and cool situations for optimum growth. **USES** Leaves are used as fodder. Soft wood is used for packing cases and matchsticks. Due to its strong root system, the tree helps to prevent soil erosion.

Indian Willow ■ *Salix tetrasperma*

Tree. Height: 7–9m

DESCRIPTION Slender, deciduous tree with a light, open crown. Simple leaves are narrow, lance shaped and arranged alternately. They are smooth and glossy above and greyish-green underneath. Pale yellow flowers are minute and grow in drooping spikes in December–March. They are followed by fruits that form capsules containing seeds with silky hair. **HABITAT** Mostly found in the plains alongside streams. Thrives in moist conditions. **USES** Elegant tree used in landscapes near water bodies. Its roots help to bind soil and prevent erosion. Delicate, flexible twigs used in basket making.

Weeping Willow ■ *Salix babylonica*

Tree. Height: up to 15m

DESCRIPTION Tree with a distinct drooping form covered with fine leaves, hence the name Weeping Willow. Long, narrow leaves are simple and shaped like a lance. Flowers are minute and clustered along a short, drooping stem. Male and female flowers appear on different trees and are pollinated by bees.
HABITAT Native to China and a tree of temperate climates. Also does well in subtropical regions if the soil is deep and damp. Prefers locations along rivers and water bodies. **USES** Planted as an ornamental specimen tree in gardens and parks. The aggressive root system prevents soil erosion along riverbanks. Medicinally, the bark and leaves are considered astringent and tonic.

Betel Nut Palm

Areca catechu

Palm. Height: up to 20m

DESCRIPTION Tall, slender palm with a solitary stem, growing straight, sometimes with a graceful bend of the trunk. Pale-coloured trunk is topped by a green crown shaft from which the leaf fronds emerge. Fronds are broad and have toothed edges. Whitish flowers appear in clusters at the base of the shaft. Oval-shaped fruits are yellow-red and contain a single seed similar to a nutmeg. It is covered by a fibrous outer layer. **HABITAT** Widely distributed, it is cultivated in many parts of the hot tropical regions of India along the coasts and inland. It prefers moist, well-drained soil and a location in partial shade. **USES** Commercially grown for the nut inside the fruit, popularly called Betel nut. It is a chewed as a stimulant and is known to be addictive and mildly narcotic.

Bismarckia Palm

■ *Bismarckia nobilis*
Palm. Height: 15–20m

DESCRIPTION Wide-spreading palm with stout trunk. Fan-shaped fronds are grey-green in colour. Each leaf can spread to up to 3m across and is divided into many segments. Leaves are nearly circular in shape and have folded pleats. Flower stalk emerges from among leaves, and male and female flowers occur on separate plants. Fruits are round, and about 4cm in diameter. Seed is the size of a walnut. **HABITAT** Originally from Madagascar, and widely grown in subtropical and tropical regions. Hardy palm that does well in both dry and moist conditions, but prefers well-drained soil. **USES** Relatively new palm in Indian gardens, where it is used as a specimen plant.

Coconut ■ *Cocos nucifera*

Palm. Height: 20–30m

DESCRIPTION Tall palm with slender trunk that often leans and curves, best known for its fruit. Long leaf fronds grow to up to 6m in length and are made up of pinnate leaflets. Flowers hang in drooping clusters at bases of leaves. Large fruit has three sides and is covered in a dense husk with a hard shell that contains the pulp and water. Green fruit ripens to yellow-brown as it matures. **HABITAT** Widely distributed in the tropics, Coconut is most common along the coast of the Indian peninsula. It also thrives in inland areas and can grow in poor soil. **USES** Coconut water found in centre of fruit is a favoured drink around the world – it is nutritious and refreshing. Pulp surrounding the water is eaten fresh, or shredded and used in various recipes. Coconut oil produced from the pulp is used for cooking, as well as in dressing the hair. Medicinal properties have been attributed to it. Leaves are woven into mats and used for thatching roofs. Trunk is often utilized in building construction.

Chinese Fan Palm

■ *Livistona chinensis*

Palm. Height: up to 10m

DESCRIPTION Medium-sized fan palm with a straight-growing trunk that is rough and textured with the bases of fallen leaf fronds. Fronds are nearly circular in shape, glossy and greyish-green. Blades are divided to about halfway with drooping segments. Flowers grow in large, bunching clusters up to 1.2m long; flowers are small and creamy. Glossy blue-green fruits are elongated or pear shaped. **HABITAT** Native to southern China and Japan. Favours subtropical conditions. Frost hardy and can adapt to temperate climate in sheltered areas. Does well in sandy soil. **USES** Planted across India as an ornamental palm in parks, and widely cultivated in gardens. Also suited to planting in containers. Fibre from leaf stalks is used to make ropes.

California Fan Palm

■ *Washingtonia filifera*
Palm. Height: up to 18m

DESCRIPTION Robust palm with a single, straight-growing trunk. Fronds spread out like fans, and are divided into many segments. Cotton-like fibre separates in strings from margins of segments. Dead fronds hang down, covering the trunks – over time, this becomes a distinct visual characteristic of the species. Small, pale yellow to cream flowers produced in dense, drooping clusters, covered by a sheath. Fruits are small black berries. **HABITAT** Native to arid regions of California and Arizona, USA, and thrives in full sun in hot, dry climates. Popular in parks and gardens almost throughout India. **USES** Planted as an ornamental.

Date Palm ■ *Phoenix sylvestris*

Palm. Height: up to 12m

DESCRIPTION Handsome palm with single trunk that grows slender and straight, and is often curved. Leaf fronds are silvery or greyish-green in colour, and leaflets end in spiky, sharp tips. Faintly scented, cream flowers are small, and produced in dense clusters. Male and female flowers grow on separate plants. Ovoid fruits hang in large clusters and ripen to orange-yellow. **HABITAT** Native to the Indian subcontinent. Grows well in sandy areas generally close to places with high water tables and water streams. Hardy plant that is drought and frost tolerant. **USES** Sap tapped from trunk is used for making jaggery (a type of unrefined cane sugar used in Asia and Africa), as well as fermented to make toddy across large parts of India. Leaves are used for weaving mats and baskets. Fruit is often utilized in traditional medicine. The species is also a popular landscape plant.

Doum Palm ■ *Hyphaene dichotoma*

Palm. Height: 10–12m

DESCRIPTION Slender palm with trunks that fork out into branches. Trunks are marked by bases of fallen leaf stalks. Leaf fronds are fan shaped and clustered at the ends of branches. Male and female plants appear in spikes on separate palms. Fruits are irregular in shape with rounded edges. They stay on the palm for a long time and are covered in a fibrous outer layer. **HABITAT** Known to have been introduced to India from Africa. Rare species not commonly seen in India. Grows in parts of Gujarat, Daman and Diu. Specimen palms in Calcutta botanical gardens thrive in that area. Grows in areas of high water tables such as banks of rivers and streams. **USES** Grown as a specimen plant in gardens. Leaves are used to weave baskets and roof thatch. Useful in checking soil erosion on river banks and streams.

Fishtail Palm ■ *Caryota urens*

Palm. Height: up to 12m

DESCRIPTION Tall, straight-growing palm with drooping fronds. These comprise leaflets that are twice compound and shaped like the tail of a fish (hence the common name of the species). Prominent drooping stems bear flowers and fruits. Flowers appear at various times of the year. When cut, stem yields a sap that is rich in sugar and vitamins. Fruits are conspicuous and round, darkening to deep red when ripe. They contain fibrous flesh that can irritate the skin on contact. **HABITAT** Native to India, Burma and parts of Malaysia. **USES** Sap from stem is drunk fresh and locally known as *Toddy*. It is also fermented to make an alcoholic drink called *Arrack*. Fibre from leaves is used to make ropes and in basket weaving.

Foxtail Palm

■ *Wodyetia bifurcata*

Palm. Height: up to 9m

DESCRIPTION Medium-sized palm with unusually lush, attractive foliage. Pale trunk with leaf scars grows straight, but becomes bulbous as it matures. As in Royal Palm (see p. 168), top of trunk is a smooth green shaft from which leaf fronds emerge. Fronds are dense and bright green, long and drooping like the tail of a fox. They comprise compound, pinnate leaves. White flowers hang in clusters from base of green shaft. Male and female flowers occur on the same palm. Oval fruits are green and ripen to orange-red. **HABITAT** Native to Australia. Can tolerate hot and dry climates, and prefers a spot in full sun. **USES** Planted as a specimen palm in Indian gardens. An uncommon plant.

Licuala Palm *Licuala spinosa*

Palm. Height: 5–6m

DESCRIPTION Medium-sized palm with clustered multiple slender stems forming large clumps. Grows to up to 5m in height. Circular-shaped leaf fronds are deeply divided into many segments. Edges of the fronds are blunt and toothed. Yellowish flowers are minute and hairy, and appear in clusters. Fruits are single-seeded green berries that ripen to orange. **HABITAT** Palm of the humid rainforests distributed in South-east Asia, particularly the Philippines and Thailand. Also found in the Andaman Islands. This species thrives is moist conditions in partial shade. **USES** Often planted as a specimen palm found in Indian gardens on the coast. Leaves are used to make thatch for covering roofs and hand-held fans.

Palmyra Palm

■ *Borassus flabellifer*
Palm. Height: up to 20m

DESCRUPTION Towering, straight-growing palm, referred to as a giant among fan-leaved palms. Fan-shaped frond is divided by deep clefts into many segments. Male and female flowers grow on separate plants, in long, drooping panicles. Fruit contains a juicy sweet pulp that is the kernel surrounding the seed. **HABITAT** Cultivated in many parts of India and South-east Asia for commercial purposes. **USES** Fruit is a delicacy and is popular in local markets. Many parts of the palm are used. Sap produced from cuts in the trunk is rich in sugar, and is used to produce jiggery. It ferments naturally to form *Toddy*, an alcoholic drink, and can also be distilled to create the liquor *Arrack*. Leaves are used to weave baskets and mats in many parts of India.

Rhapis Palm ■ *Rhapis excelsa*

Palm. Height: 4.5m

DESCRIPTION Low-height palm forming clumps of slender, multiple stems. Fan-shaped fronds are divided into 5–8 segments. They are blunt at the tips, unlike those of the **Slender Lady Palm** *Rhapis humilis*, which looks similar but has fronds with pointed ends. Creamy-white flowers are minute and inconspicuous. Fruits are small berries containing a single seed. **HABITAT** Introduced to India, and native to south-east China and Taiwan. Popular in Indian gardens, thriving in moist growing conditions in partial shade. **USES** Grown for its dense green foliage as an ornamental in gardens. Dense clumps of the plant grow straight and can be closely planted as hedges.

Royal Palm ■ *Roystonea regia*

Palm. Height: up to 30m

DESCRIPTION Tall, stately palm with pale, straight-growing trunk. Some trunks develop a slightly bulbous form over time. Top of trunk tapers to green stem, from which stems of leaf fronds cascade out and spread. Fronds can grow to up to 3–4m long, and comprise long, narrow, bright green pinnate leaflets. Whitish, pale yellow flowers appear in dense clusters on drooping spikes. Fruits are purple berries. **HABITAT** Common and popular ornamental palm, originating on the islands of the Caribbean and widely cultivated in India. Thrives in full sun in hot tropical climates. **USES** Often planted in straight lines in gardens and parks to emphasize its formality.

Sealing Wax Palm ■ *Cyrtostachys renda*

Palm. Height: 6–10m

DESCRIPTION Slender palm that grows tall and has multiple clumping stems. Colour of stems is deep red, matching that of the wax used in seals, hence the common name of the species. Bright green leaf fronds are pinnate and form graceful arches. Flowers appear in branched clusters at bases of leaves. Small fruits are elongated, almost elliptical in shape and each contains a single seed. **HABITAT** Coastal palm from Thailand, Malaysia and Indonesia. Grown as an ornamental in hot, humid tropical regions of India such as Kerala in south and Kolkata in east. **USES** Planted as an ornamental. Due to clumping habit, also suited to planting in a line as a screen.

Japanese Sago Palm ■ *Cycas revoluta*

Cycad. Height: up to 3m

DESCRIPTION Short, single trunk topped by rosettes of fronds that have stiff, needle-like, leathery leaves. Male cones are cylindrical. **HABITAT** Native to southern islands of Japan and popularly cultivated in gardens throughout India. Thrives in full sun. **USES** An ornamental plant for gardens. Seeds and tubers contain starch and can be cooked and eaten.

Queen Sago Palm ■ *Cycas circinallis*

Cycad. Height: up to 4.5m

DESCRIPTION Graceful leaves are drooping and not stiff. Multiple trunks are developed on mature plants. Seeds are yellow-red in colour. **HABITAT** Native to South India, Sri Lanka and other parts of South-east Asia. Thrives in warm tropical climate and favours a location in partial shade. **USES** Sago, an edible starch, is extracted from the trunk. Seeds are also rich in starch and are eaten at times of food scarcity. *Cycas* is an ancient plant genus comprising 60 species distributed across Australia, South-east Asia, Madagascar and East Africa. They have a palm-like structure, with a central trunk topped by fronds radiating in spirals from the top. The fronds are made up of pinnate leaves that are stiff and narrow. Male and female cones appear on separate plants. Two of the most popular *Cycas* species seen in India are *Cycas circinallis* and *C. revoluta.* They are very slow growing and have a lifespan of 50–100 years.

Further Reading

Krishen, P. 2006. *Trees of Delhi A Field Guide*. Dorling Kindersley
Krishen, P. 2013. *Jungle Trees of Central India*. Penguin Books India
Sahni, K. C. 1998. *The Book of Indian Trees*. Bombay Natural History Society
Jain S. K. 1968. *Medicinal Plants*. National Book Trust India
Blatter E. & Millard W. S. 1937. *Some Beautiful Indian Trees*. Bombay Natural History Society
Bose, T. K. & Chowdhury B. 1991. *Tropical Garden Plants in Colour*. Horticulture and Allied Publishers
Barwick M. 2004. *Tropical and Subtropical Trees A Worldwide Encyclopaedic Guide*. Thames and Hudson
Patnaik N. 1993. *The Garden of Life*. Doubleday
Chakravarty V. 1976. *Our Tree Neighbours*. National Council of Educational Research & Training
Cowen D. V. 1950. *Flowering Trees & Shrubs in India*. Thacker and Co. Ltd.
Krishna N. & M. Amirthalingam. 2014. *Sacred Plants of India*. Penguin Books
Harrison L. 2012. *Latin for Gardeners*. The University of Chicago Press

Websites

www.worldagroforestry.org
www.indiabiodiversity.org
www.flowersofindia.net
www.efloraindia.nic.in
www.jntbgri.in
www.kew.org

ACKNOWLEDGEMENTS

The authors would like to thank their colleagues at PSDA studio, New Delhi for taking on extra responsibility at work while the book was being researched. In particular, Madhu Shankar, Vishwesh Vishwanathan, Arti Mathur, Shivani Agarwal, Ayush Sitholay and Kriti Dhingra. Center for Science and Environment led by Sunita Narain shared additional photographs and information on sacred groves of India. Nicholas Vreeland, Pradip Krishen, Mimi Roberts, Alpana Khare and Gowri Mohanakrishnan also generously contributed pictures. Nicholas Vreeland also gave valuable tips on photography.

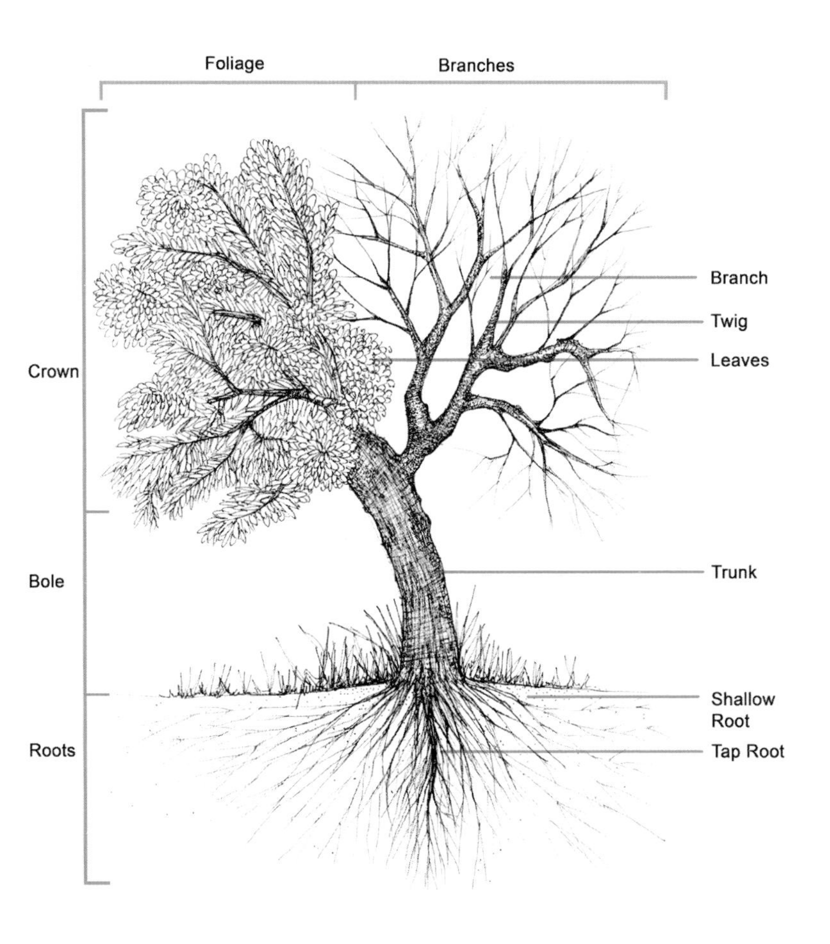
Foliage
Branches
Crown
Bole
Roots
Branch
Twig
Leaves
Trunk
Shallow Root
Tap Root